SPRINGER PROCEEDINGS IN PHYSICS 113

SPRINGER PROCEEDINGS IN PHYSICS

91 **The Dense Interstellar Medium in Galaxies**
Editors: S. Pfalzner, C. Kramer, C. Straubmeier, and A. Heithausen

92 **Beyond the Standard Model 2003**
Editor: H.V. Klapdor-Kleingrothaus

93 **ISSMGE**
Experimental Studies
Editor: T. Schanz

94 **ISSMGE**
Numerical and Theoretical Approaches
Editor: T. Schanz

95 **Computer Simulation Studies in Condensed-Matter Physics XVI**
Editors: D.P. Landau, S.P. Lewis, and H.-B. Schüttler

96 **Electromagnetics in a Complex World**
Editors: I.M. Pinto, V. Galdi, and L.B. Felsen

97 **Fields, Networks, Computational Methods and Systems in Modern Electrodynamics**
A Tribute to Leopold B. Felsen
Editors: P. Russer and M. Mongiardo

98 **Particle Physics and the Universe**
Proceedings of the 9th Adriatic Meeting, Sept. 2003, Dubrovnik
Editors: J. Trampetić and J. Wess

99 **Cosmic Explosions**
On the 10th Anniversary of SN1993J (IAU Colloquium 192)
Editors: J. M. Marcaide and K. W. Weiler

100 **Lasers in the Conservation of Artworks**
LACONA V Proceedings, Osnabrück, Germany, Sept. 15–18, 2003
Editors: K. Dickmann, C. Fotakis, and J.F. Asmus

101 **Progress in Turbulence**
Editors: J. Peinke, A. Kittel, S. Barth, and M. Oberlack

102 **Adaptive Optics for Industry and Medicine**
Proceedings of the 4th International Workshop
Editor: U. Wittrock

103 **Computer Simulation Studies in Condensed-Matter Physics XVII**
Editors: D.P. Landau, S.P. Lewis, and H.-B. Schüttler

104 **Complex Computing-Networks**
Brain-like and Wave-oriented Electrodynamic Algorithms
Editors: I.C. Göknar and L. Sevgi

105 **Computer Simulation Studies in Condensed-Matter Physics XVIII**
Editors: D.P. Landau, S.P. Lewis, and H.-B. Schüttler

106 **Modern Trends in Geomechanics**
Editors: W. Wu and H.S. Yu

107 **Microscopy of Semiconducting Materials**
Proceedings of the 14th Conference, April 11–14, 2005, Oxford, UK
Editors: A.G. Cullis and J.L. Hutchison

108 **Hadron Collider Physics 2005**
Proceedings of the 1st Hadron Collider Physics Symposium, Les Diablerets, Switzerland, July 4–9, 2005
Editors: M. Campanelli, A. Clark, and X. Wu

109 **Progress in Turbulence 2**
Proceedings of the iTi Conference in Turbulence 2005
Editors: M. Oberlack et al.

110 **Nonequilibrium Carrier Dynamics in Semiconductors**
Proceedings of the 14th International Conference, July 25-29, 2005, Chicago, USA
Editors: M. Saraniti, U. Ravaioli

111 **Vibration Problems ICOVP 2005**
Editors: E. Inan, A. Kiris

112 **Experimental Unsaturated Soil Mechanics**
Editor: T. Schanz

113 **Theoretical and Numerical Unsaturated Soil Mechanics**
Editor: T. Schanz

Volumes 66–90 are listed at the end of the book.

T. Schanz (Ed.)

Theoretical and Numerical Unsaturated Soil Mechanics

With 111 Figures and 21 Tables

Springer

Professor Dr. Ing. habil. Tom Schanz
Bauhaus-Universität Weimar
Laboratory of Soil Mechanics
Coudraystrasse 11c
99421 Weimar, Germany
tom.schanz@bauing.uni-weimar.de

ISSN 0930-8989

ISBN 978-3-540-69875-3 Springer Berlin Heidelberg New York

Library of Congress Control Number: 2007921084

Springer is a part of Springer Science+Business Media
springer.com

Typesetting: Integra Software Services Pvt. Ltd., Pondicherry, India
Cover design: WMX-Design, Heidelberg

Printed on acid-free paper SPIN: 11977490 42/Integra 5 4 3 2 1 0

Foreword

The event is a continuation of the series of International Conferences on Unsaturated Soils in Germany. The first International Conference was held during September 2003 in Bauhaus-University Weimar, Weimar, Germany. The current event is the second one in the series entitled "Mechanics of Unsaturated Soils." The primary objective of the Conference has been to discuss and understand unsaturated soil behaviour such that engineered activities are made better with times in terms of judgement and quality. We all realise by now that in addition to the knowledge on the classical concepts, it becomes an enormous challenging task to adapt convincing new concepts and present them in such a way that it could be used in engineering practices. During the last six years or so (2001–2007), scientific research works were extensively taken up by five scientific research teams from five German universities, whose scientific leaders are Wolfgang Ehlers (Universität Stuttgart), Jens Engel (HTW Dresden), Rainer Helmig and Holger Claas (Universität Stuttgart), Tom Schanz (Bauhaus Universität Weimar), Christos Vrettos, Helmut Meissner and Andreas Becker (Universität Kaiserslautern). The research studies involved theoretical and numerical approaches along with experimental studies on unsaturated soils. These two volumes present recent research findings obtained within this collaboration by the above research groups along with excellent contributions from several research groups throughout the World.

The experimental studies reported herein primarily focussed on the role of microstructure and fabric for the complex coupled hydro-mechanical behaviour of cohesive frictional materials. Several papers considered the relevance of temperature affecting the constitutive behaviour of clays. A careful reader may recognise that in both the topics there is an ambiguity with regard to the conclusions derived. Common features of state of the art theoretical and numerical approaches, including TPM (theory of porous media) and mixture theory, intend to describe the complex multi-field problems of fully coupled thermo-hydraulic-mechanical-chemical initial-boundary value problems. Additional important field of research includes optimization of numerical schemes to gain better computational performance. Applications include highly toxic waste disposals, slope stability problems and contaminants transport in porous media. Some major significant contributions from the invited and keynote speakers are also included.

I would like to extend my deep sense of appreciation as the editor and the Head of the organizing committee, to many persons who have contributed either directly or indirectly to organize the International conference and to finalize these lecture notes. I would like to congratulate the authors for their very interesting presentations and the reported results and advances in the topics of the conference. I would like to thank all of those who promoted the conference in their respective home countries. These two volumes would have been not possible without financial support by the German Research Foundation (DFG, Deutsche Forschungsgemeinschaft) through grant FOR 440/2. We gratefully acknowledge the support of ISSMGE, especially TC6 "Unsaturated Soils" with its chairman Eduardo Alonso. I appreciate the effort of the members of the Technical committee and reviewers, who have spent their time to select the valuable contributions and to suggest the changes improving the presentation of the submitted papers. Finally, I wish to convey my thanks to all the keynote and invited speakers, authors, and delegates attending the conference.

I would like to express my deep sense of gratitude for the outstanding work performed by those involved in the technical and administrative organization of these proceedings. Special thanks go to Yvonne Lins. Typesetting of the proceedings was done by Venelin Chernogorov (alias Wily, Sofia University, Bulgaria, `wily@fmi.uni-sofia.bg`) in cooperation with Maria Datcheva. Last but not least we appreciate the fruitful cooperation with Springer publishers, especially the guidance provided by Thomas Ditzinger.

Weimar,
March 2007

Tom Schanz

Contents

Part I Constitutive Modelling

Part I

Constitutive Modelling

Mechanical Model for Unsaturated MX–80

Mattias Åkesson and Harald Hökmark

Clay Technology AB, Ideon, S-223 70 Lund, Sweden
ma@claytech.se, hh@claytech.se

Summary. In this paper a new elastoplastic model for unsaturated swelling clay is presented. A dual porosity framework is chosen, so that the void ratio is the sum of a water-filled micro void ratio, and an air-filled macro void ratio. The model is formulated as two differential equations, one each for elastic and plastic strains respectively. Each equation describes a relation between the macro void ratio and the two independent variables stress and micro void ratio. The established concept of swelling pressure and its void ratio dependence is incorporated in the model. To do this, the grain-to-grain contact stress and the contact area had to be considered.

In its present form the model only handles one-dimensional problems, e.g. compression tests with uniaxial strain and water uptake test at constant volume with assumed isotropic conditions. In order to match experimental results from these two type of tests, it was found to be necessary to modify the parameter value regarding the contact area. This difference was analyzed and justified through a geometrical investigation. This study also indicated that uniaxially compacted samples are characterized by an anisotropic grain structure.

Key words: elastoplastic model, swelling pressure, contact stress, unsaturated clay, bentonite, dual porosity

1 Introduction

Knowledge of the mechanical behavior of compacted bentonite is essential for the understanding of barrier processes in KBS–3 type nuclear waste disposal systems (SKB 1999). A consistent constitutive model is necessary for numerical analysis of such systems.

Two codes, `Abaqus` and `Code_Bright`, are used within the Swedish nuclear waste disposal program for THM modeling of buffer and backfill materials. The latter code allows for a physically more realistic representation of various THM aspects, and may be the main tool used within the Swedish program for some applications in the future.

The thermoelastoplastic constitutive laws used in `Code_Bright` are based on the Barcelona Basic Model (BBM). BBM has, however, certain limitations for expansive soils, and has therefore been further developed into the Barcelona Expansive Model (BExM) by Alonso et al. (1999).

The model presented in this paper represents an alternative approach to describe the mechanical behavior of swelling clays, such as the buffer materials considered for the KBS–3 repository. The basic assumptions, the mathematical formulation and the parameter settings are first described. Application examples for compression tests at constant water content and water uptake tests at constant volume are thereafter presented. Finally, the need for and use of different parameter values for the contact area function α for simulating the two types tests are analyzed and discussed.

2 Framework for a New Model

The model is based on the following assumptions:

i. The porosity is two-parted in micro- and macro-porosity. The void ratio e is thus the sum of a micro (e_m) and a macro (e_M) void ratio. The same approach is used in BExM (Alonso et al. 1999).
ii. The micro-porosity is water-filled, whereas the macro-porosity is air-filled. The micro void ratio is thus equal to the volumetric water content, i.e. $e_m = w \cdot \rho_s/\rho_w$, where w is the gravimetric water content and ρ_s/ρ_w is the solid-water density ratio (= 2.78).
iii. The two void ratios (e_m and e_M) and the stress (σ), are the state variables of the model (σ is positive for compressive stresses).
iv. The contact stress (σ') between grains is defined by a ratio, $\alpha = \sigma/\sigma'$, corresponding to the ratio between the contact area and the section area. α is assumed to be a function of e_m and e_M.
v. The swelling pressure (p_s) is a central quantity of the model, and is determined by the micro void ratio only.
vi. A condition for elastic strains is that $\sigma' < p_s$, while $\sigma' = p_s$ for plastic strains. A transition from elastic to plastic strains is made when the latter condition is fulfilled.
vii. The elastic domain is governed by a modulus of compression (K), determined by the micro void ratio only.

3 Model Description

The elastic compression modulus is defined as $\mathrm{d}\varepsilon = -\mathrm{d}\sigma'/K$, where $\mathrm{d}\varepsilon$ is the volumetric strain, which is further expressed as $\mathrm{d}e_M/(1 + e_m + e_M)$. $\mathrm{d}\sigma'$ can be developed in terms of $\mathrm{d}\sigma$, $\mathrm{d}e_m$ and $\mathrm{d}e_M$:

$$\mathrm{d}\left(\frac{\sigma}{\alpha}\right) = \left(\frac{\mathrm{d}\sigma}{\alpha} - \frac{\sigma}{\alpha^2} \cdot \mathrm{d}\alpha\right) = \frac{\mathrm{d}\sigma}{\alpha} - \frac{\sigma}{\alpha^2} \cdot \left[\frac{\partial \alpha}{\partial e_m} \cdot \mathrm{d}e_m + \frac{\partial \alpha}{\partial e_M} \cdot \mathrm{d}e_M\right] . \quad (1)$$

Taken together, the elastic relation can be expressed as:

$$\left[\frac{\sigma}{\alpha} \cdot \frac{\partial \alpha}{\partial e_M} - \frac{\alpha \cdot K}{1 + e_m + e_M}\right] \cdot \mathrm{d}e_M = \mathrm{d}\sigma - \frac{\sigma}{\alpha} \cdot \frac{\partial \alpha}{\partial e_m} \cdot \mathrm{d}e_m . \quad (2)$$

According to the condition for plastic strains, changes in contact stress equal changes in swelling pressure. The change in swelling pressure can be expressed in terms of changes in the micro void ratio e_m:

$$\mathrm{d}p_s = \frac{\mathrm{d}p_s}{\mathrm{d}e_m} \cdot \mathrm{d}e_m . \quad (3)$$

Together with Eq. (1), the plastic relation can thus be expressed as:

$$\frac{\sigma}{\alpha} \cdot \frac{\partial \alpha}{\partial e_M} \cdot \mathrm{d}e_M = \mathrm{d}\sigma - \left[\alpha \cdot \frac{\mathrm{d}p_s}{\mathrm{d}e_m} + \frac{\sigma}{\alpha} \cdot \frac{\partial \alpha}{\partial e_m}\right] \cdot \mathrm{d}e_m . \quad (4)$$

A successive transition between elastic and plastic strains can also be formulated. With this option it is possible to generate developments, e.g. stress paths, without the abrupt changes in slope that follow from a discrete transition from elastic to plastic strains. It is, however, beyond the scope of this paper to include this option.

4 Parameter Settings

The functions used in the model for the modulus of compression, swelling pressure and the contact area, respectively, are given as follows:

$$K(e_m) = K_1 \cdot \mathrm{e}^{5\cdot(1-e_m)} \qquad (\mathrm{MPa}) , \quad (5)$$

$$p_s(e_m) = 2 \cdot e_m^{\frac{-1}{0.24}} \qquad (\mathrm{MPa}) , \quad (6)$$

$$\alpha(e_m, e_M) = \left(\frac{1 + e_m}{1 + e_m + e_M}\right)^{\gamma} \qquad (-) . \quad (7)$$

Eq. (5) gives the value K_1 for the modulus of compression at a micro void ratio of 1. The K_1 value for 1D compression tests is 100 MPa, while it is 50 MPa for isotropic 3D test. This difference in modulus for the two different test conditions can be compared with the ratio $3(1-\nu)/(1+\nu)$ between the oedemeter- and bulk moduli (Jaeger 1956). With a value of Poisson's ratio ν of about 0.15 for compacted MX–80 with a water content of 12% (Kalbantner and Johannesson 2000), that ratio is about 2.

The swelling pressure curve follows the one presented by Börgesson at al. (1995), but with a modified exponent, which was changed from $-1/0.187$ to $-1/0.24$.

The contact area function is based on the ratio between the specific grain volume $(1 + e_m)$ and the total specific volume $(1 + e_m + e_M)$. The exponent γ is set to 7 and 15, for 1D compression tests and isotropic 3D tests, respectively. Geometrical investigations, presented below in Sect. 6, indicate that this relation describes the contact area of reasonably realistic grain structures well.

5 Examples of Model Results for Different Test Conditions

The model has been tested for two different test conditions: compression at constant water content ($de_m = 0$) and water uptake during constant volume ($de_m + de_M = 0$). The differential equations for these test conditions can easily be derived from Eqs (2) and (4) through ignoring one of the terms or through rearrangement of terms. The differential equations were solved through simple explicit stepping using the programming options of the MathCad® software.

The compression tests used for calibration and comparisons were constant rate of strain tests (CRS), in which load and deformation was measured continuously on pre-compacted samples (Börgesson 2001). Experimental results from three of these tests are shown in Fig. 1 together with model calculations. The model captures the elastic and the plastic slopes, as well as the apparent pre-consolidation pressures with good accuracy.

The constant volume tests used for comparison were performed on series of compacted samples in oedometers (Börgesson 1985). Each sample was allowed to equilibrate at an elevated water content before the related pressure was registered. Experimental results from one of these tests are shown in Fig. 2 together with model calculations. The model is designed to end at a swelling

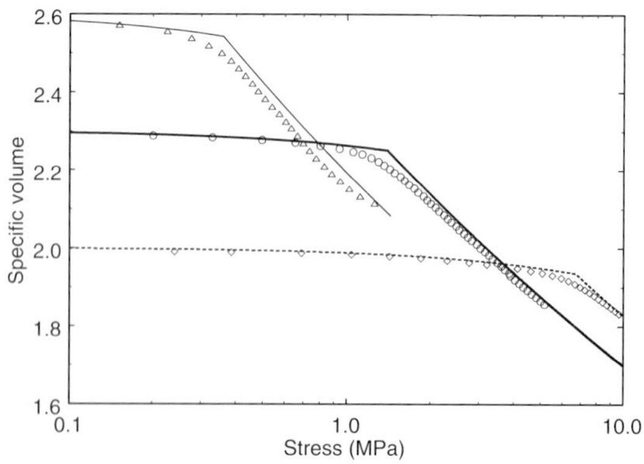

Fig. 1. Test results from CRS test (symbols, data from Börgesson 2001) and model calculation (lines) of specific volume vs. stress at constant water content. Water contents were 17% ◇, 23% o and 39% △

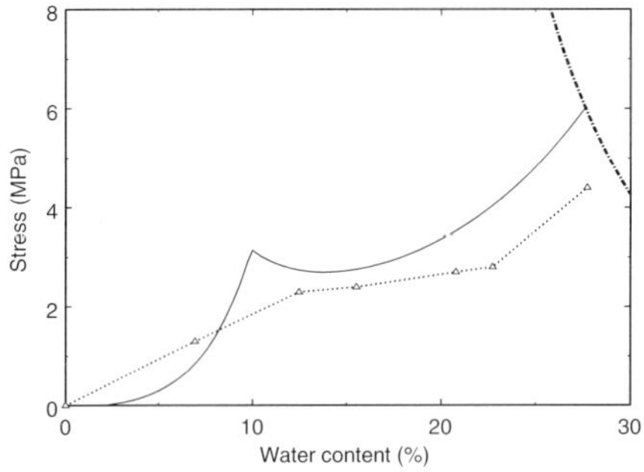

Fig. 2. Test results from constant volume test (symbols, data from Börgesson 1985) and model calculation (line) of σ vs. w. Bold line to the upper right is the p_s-curve. The void ratio was 0.77; the initial water content was set to 2%

pressure given by the p_s-function, i.e. Eq. (6). During plastic conditions, the model generates a certain plateau, which is found in the experimental results as well.

The model gives a stress peak at the transition from elastic to plastic conditions. This can be smoothen by using the successive transition mentioned at the end of Sect. 3. The peak would be significantly higher (even exceeding the final swelling pressure) if the γ value for 1D compression test was used. This was the main reason for considering an alternative value for the 3D constant volume case. The higher K_1 value used for 1D test would in a similar way result in a steeper stress buildup during the elastic phase.

It can be mentioned that experimental results from constant load tests ($\mathrm{d}\sigma = 0$) have been compared with the model as well. It has been found that the model, with the parameter setting used for the constant volume tests, can mimic experimental results with approximately the same accuracy as for the constant volume tests.

6 Comments on the Contact Area Function

The differences between the contact area functions given for different test conditions can be justified by analyzing a schematic grain structure (Fig. 3). The basis for this approach is to generalize the specific volume to a certain cell geometry. The cell geometry chosen here is a truncated octahedron described by a set of lengths (l_x, l_y, l_z). This form can be tightly packed and can easily be described by three such lengths. In a similar way, the specific grain volume is described by another set of lengths ($r_x l_x$, $r_y l_y$, $r_z l_z$), which represents the radii of an ellipsoid truncated by the current cell geometry.

Grains are in contact at points at the cell surface where the ellipsoid is truncated. The α-function corresponds to the ratio between the contact area,

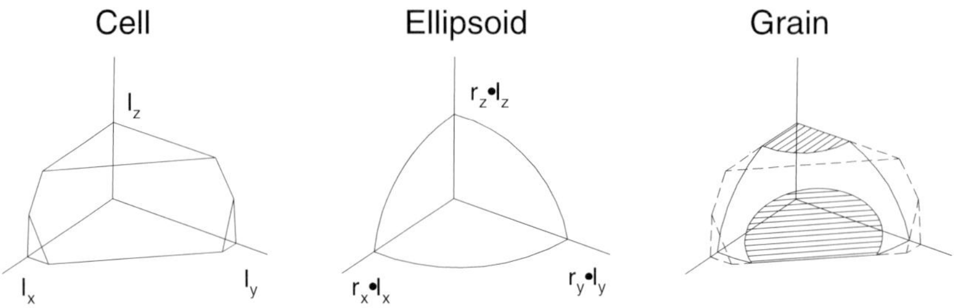

Fig. 3. Definition of cell and grain geometry. Only one eighth of the geometry is shown. The hexagonal face of the cell is a symmetry plane between the origin and (l_x, l_y, l_z). The marked areas of the grain are contact areas

projected on the x–y plane, and the base area of the cell in that same plane, versus the ratio between the grain and cell volumes, respectively.

In this structure, the 1D compression tests with constant w correspond to a condition with constant and equal radial cell lengths ($l_x = l_y$) and a constant grain volume. The cell volume is thus proportional to the axial cell length (l_z). Even if the grain volume is constant and the conditions are axis-symmetric ($r_x = r_y$) it can be noted that the grain geometry exhibits a certain degree of freedom so that different sets of r_x and r_z can be found. For reproduction of experimental results a tentative rule has been found, according to which r_z is 20% higher than r_x. Moreover, a slightly initially compressed cell structure ($l_z/l_x = 0.7$) is required for an initial $V_{\mathrm{grain}}/V_{\mathrm{cell}}$-ratio of 0.7. Results from this calculation are shown in Fig. 4.

The 3D constant volume test with increasing water content is equivalent to a cell geometry with fixed values of l_x, l_y, and l_z. For changing the water content the characteristic length ratios of the grain (r_x, r_y and r_z) can be varied. As a tentative approach these ratios were given the same value. In order to mimic the low contact area suggested by experimental results, it was necessary to choose a considerably compressed cell structure ($l_z/l_x = 0.3$). Results are shown in Fig. 4. Such a structure may be expected given that the samples were compressed uniaxially.

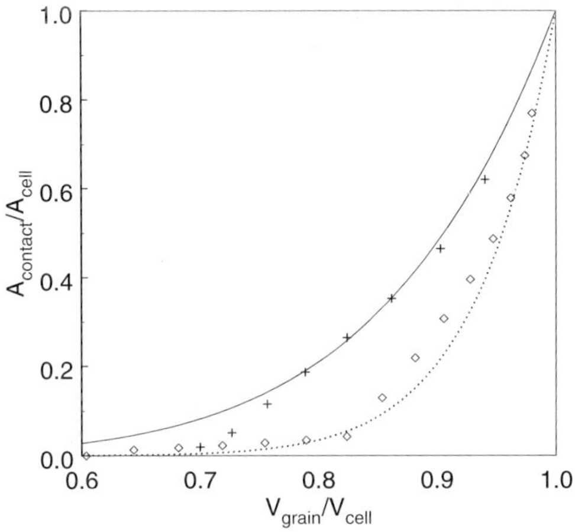

Fig. 4. Comparison of α-functions ($\gamma = 7$ solid line; $\gamma = 15$ dotted line) with corresponding geometric relations (constant water content $+$; constant volume $\diamondsuit$)

7 Concluding Remarks

The presented model is found to capture several characteristics of compacted MX–80 in a consistent way. It should, however, be observed that the chosen parameter functions, especially the chosen swelling pressure curve, probably has certain limitations. The compression tests used for the calibration were performed on samples with water contents not lower than 17%. Modeling of samples with significantly lower water contents (which was made for the constant volume tests) should therefore be made with caution, unless new information can support the current function.

The indications of an anisotropic grain structure in uniaxially compacted samples may also have important consequences. The possible differences between uniaxially and isostatically compacted samples should be verified through experimental work before any further conclusions are drawn.

Finally, it should be clear that the model needs to be adapted to general 3D problems in order to be fully usable for numerical analysis. The geometry outlined in Fig. 3 could possibly be an adequate framework for such a generalization. Further work is, however, needed before this can be realized.

Acknowledgements

The authors wish to acknowledge that this paper is a result of work funded by the Swedish Nuclear Fuel and Waste Management Company (SKB).

References

Alonso EE, Vaunat J, Gens A (1999) Modelling the mechanical behaviour of expansive clays, Eng Geol 54:173–183
Börgesson L (1985) Water flow and swelling pressure in non-saturated bentonite-based clay barriers, Eng Geol 21:229–237
Börgesson L, Johannesson L-E, Sandén T, Hernelind J (1995) Modelling of the physical behaviour of water saturated clay barriers. Laboratory tests, material models and finite element application. SKB TR 95–20
Börgesson L (2001) Compilation of laboratory data for buffer and backfill materials in the Prototype Repository, SKB IPR–01–34
Jaeger JC (1956) Elasticity, Fracture and Flow. Methuen & Co. Ltd, London
Kalbantner P, Johannesson L-E (2000) Hållfasthetsberäkningar för en bentonitbuffert bestående av enaxligt kompakterade bentonitkroppar, SKB R–00–42
SKB (1999) SR 97 – Post-closure safety, SKB TR–99–06

Hypoplastic Modeling of the Disintegration of Dry and Saturated Weathered Broken Rock

Erich Bauer[1], Sendy Fransiscus Tantono[1], Weijun Cen[2], Yueming Zhu[2], and Karl Kast[3]

[1] Institute of Applied Mechanics, Graz University of Technology, Graz, Austria
erich.bauer@tugraz.at, sendy.tantono@gmail.com
[2] Institute of Hydraulic Structural Eng., Hohai University, Nanjing, P.R. China
cenweijun@163.com, hhuzym@126.com
[3] Consulting Geotechnical Eng. ICOLD, Germany karl.kast@t-online.de

Summary. For weathered broken rock materials the resistance of particles against abrasion and breakage is strongly influenced by their moisture content. The focus of the present paper is on the constitutive modeling of the mechanical behaviour of weathered granular materials under dry and wet conditions using a moisture dependent granular hardness. It will be shown that the results obtained from numerical simulations of element tests are in a good agreement with laboratory experiments carried out with weathered granite.

Key words: weathered grains, moisture dependent granular hardness, hypoplastic model

1 Introduction

The mechanical behavior of broken rock is different for unweathered or weathered grains. The degree of geological disintegration, i.e. by chemical weathering or by the intensity and the orientation of micro-cracks, has a significant influence on the granular hardness and as a consequence on the evolution of grain abrasion and grain breakage. Depending on the state of weathering the propagation of micro-cracks due to water-induced stress corrosion can be strongly influenced by the moisture content of the grain material. Under higher stress levels the disintegration of grains can be accelerated by moisture, which leads to a reduction of the resistance to compaction and shearing.

The focus of the present paper is on modeling the essential mechanical properties of weathered broken rock materials using a hypoplastic continuum approach. In hypoplasticity the evolution equation for the effective stress is formulated with a nonlinear isotropic tensor-valued function depending on the current state quantities and the rate of deformation. Unlike the classical concept of elasto-plasticity no decomposition of the deformation into elastic and plastic parts is needed in hypoplasticity (Darve 1991, Kolymbas 1991). In order to model unelastic material properties the rate of deformation tensor is incorporated in the constitutive equation in a nonlinear formulation. With a pressure dependent density factor the influence of pressure and density on the incremental stiffness, the peak friction angle and the dilatancy can be modeled for an initially loose or dense state using a single set of constants (Wu and Bauer 1993, Wu et al. 1996, Herle and Gudehus 1999). Limit states or so-called critical states are included in the constitutive equation for a simultaneous vanishing of the stress rate and volume strain rate (Bauer 2000). While the hypoplastic model for granular materials with unweathered grains by Gudehus (1996) and Bauer (1996) assumes a constant granular hardness, the present paper discusses an extension of this version to weathered broken rock materials using a granular hardness depending on the moisture content of the grains. In this context it is important to distinguish between the moisture content of the weathered grains and the water content of the void space between the grains. Consequently suction of the grain assembly is not a suitable quantity to describe the wet state within the fractured grains. This quantity is therefore not used to model the degradation of the granular hardness. In the present paper it will be shown that an increase of the compressibility and a decrease of the limit void ratios with an increase of the moisture content of the grains is modeled in a simplified manner using a moisture content dependent granular hardness. Due to the lack of experimental data for partly saturated states of the grains only the properties of the dry grains and water saturated grains under drained conditions are compared. The results obtained from numerical simulations of triaxial compression tests are also compared with experiments carried out with weathered granite.

2 Granular Hardness and Limit Void Ratios

For weathered rockfill materials the compressibility is higher for a wet than for a dry material as illustrated in Fig. 1. For a pre-compressed material under dry conditions (path A–B) a following wetting leads to an additional settlement (path B–C). For a continuing loading the load-displacement curve (path C–D) follows the curve A–D obtained for an initially wet material, i.e. the memory of the material of the pre-compaction under dry conditions is swept out if the load-displacement curve obtained for the wet material (path A–D) starts from the same initial density.

In the following the compression behavior is first discussed for a dry granular material and modeled using a relation between the void ratio e and the mean effective pressure $p = -(\sigma_{11} + \sigma_{22} + \sigma_{33})/3$ of the grain skeleton. The evaluation of numerous tests has shown that the compression behavior of various cohesionless granular materials can be approximated with the following exponential function:

$$e = e_0 \exp[-(3p/h_s)^n] \,. \tag{1}$$

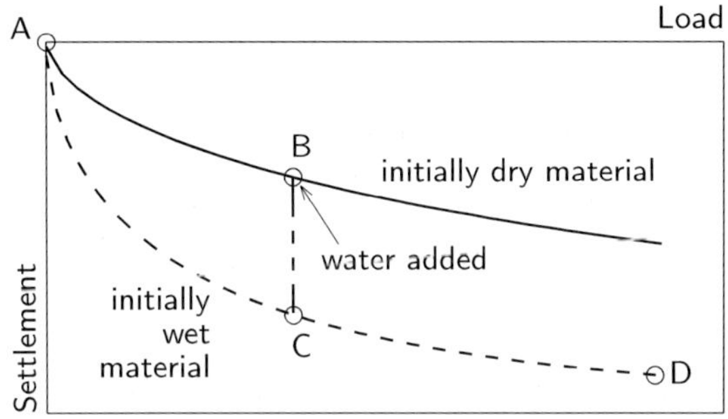

Fig. 1. Compression behavior of weathered broken rock in dry and wet states

Herein the constant e_0 denotes the void ratio for $p \approx 0$, h_s has the dimension of stress and n is a dimensionless constant. The quantity h_s is called granular hardness (Gudehus 1996) which is related to the grain aggregate under isotropic compression and different from the hardness of an individual grain. Experimental investigations show that the quantity h_s reflects the isotropic pressure where grain crushing becomes dominant. More precisely, h_s represents the isotropic pressure $3p$ at which the compression curve in a semi-logarithmic representation shows the point of inflection while the exponent n is related to the inclination of the corresponding tangent (Fig. 2(**a**)). For high pressures the void ratio in (1) tends to zero, which can be explained by grain plastification and grain crushing.

In order to model the influence of disintegration of a stressed rockfill by a reaction with water a degradation of the granular hardness with an increase of the moisture content w of the solid material is assumed in the following. To this end the constant granular hardness h_s in (1) is replaced by the moisture dependent quantity h_s^*, i.e. (Bauer and Zhu 2004)

$$h_s^* = h_{so}\psi(w)\,. \tag{2}$$

Herein h_{so} is the value of the granular hardness obtained for the dry material, i.e. h_{so} is related to $\psi = 1$, and $\psi(w) \leq 1$ denotes the disintegration factor depending on the moisture content w of the grain material. A higher moisture content means a lower value of $\psi(w)$ and of h_s^* and consequently a higher compressibility of the material as illustrated in Fig. 2(**a**). A time dependent behavior of the degradation of the granular hardness can be accounted for in a simple manner by extending relation (2), which, however, is not considered in the present paper.

It is experimentally evident that under the same pressure cohesionless granular materials can show different packing densities of the grain assembly so that the void ratio can range between a maximum void ratio e_i and a minimum void ratio e_d as shown in Fig. 2(**b**) in the so-called phase diagram of grain skeletons (Gudehus 1997). Herein the limit void ratios e_i and e_d are pressure dependent and they decrease with an increase of the mean effective

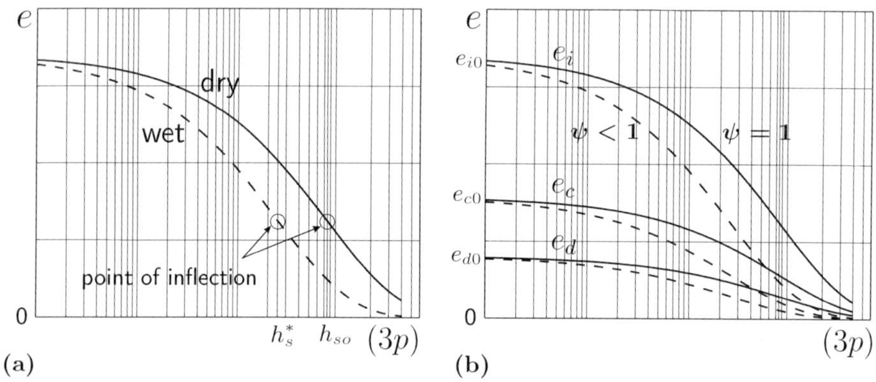

Fig. 2. Influence of the moisture content on: (**a**) compression behavior, (**b**) limit void ratios (*solid curves*: *dry state*, *dashed curves*: *wet state*)

pressure p. The upper bound, e_i, can be related to an isotropic compression starting from the loosest possible skeleton with grain contacts, i.e. there exists no homogeneous deformation which goes beyond $e = e_i$. Values of e_d will be achieved by cyclic shearing with very low amplitudes and nearly fixed mean pressure. By contrast, large monotonic shearing leads to a stationary state, which is characterized by a constant stress and constant void ratio. The void ratio in such a limit state, which is called critical void ratio, e_c, is again a pressure dependent quantity. It was suggested by Gudehus (1996) to postulate that the maximum void ratio e_i, the minimum void ratio e_d and the critical void ratio e_c decrease with the mean pressure according to

$$e_i/e_{io} = e_d/e_{do} = e_c/e_{co} = \exp[-(3p/h_s^*)^n] \,, \tag{3}$$

where e_{io}, e_{do} and e_{co} are the corresponding values for $p \approx 0$ as shown in Fig. 2(**b**). It is obvious that with a degradation of the granular hardness the pressure dependent limit void ratios and the critical void ratio are lower for $\psi(w) < 1$ as illustrated by the dashed curves in Fig. 2(**b**).

3 The Hypoplastic Model

In order to model the dependence of the incremental stiffness of weathered broken rock on the current void ratio e, the effective stress state σ_{ij}, the rate of deformation $\dot{\epsilon}_{ij}$ and the moisture dependent granular hardness h_s^* the hypoplastic constitutive model by Gudehus (1996) and Bauer (1996) is extended by the moisture content w as an additional state quantity. The proposed constitutive equations for modeling non-linear and inelastic behavior read:

$$\mathring{\sigma}_{ij} = f_s \left[\hat{a}^2 \dot{\varepsilon}_{ij} + (\hat{\sigma}_{kl}\dot{\varepsilon}_{kl})\hat{\sigma}_{ij} + f_d \hat{a}(\hat{\sigma}_{ij} + \hat{\sigma}^*_{ij})\sqrt{\dot{\varepsilon}_{kl}\dot{\varepsilon}_{kl}}\right] \,, \tag{4}$$

and

$$\dot{e} = (1 + e)\dot{\varepsilon}_v \, . \tag{5}$$

Herein $\mathring{\sigma}_{ij}$ are the components of the effective objective stress rate, $\dot{e}$ is the rate of void ratio, $\hat{\sigma}_{ij} = \sigma_{ij}/(-3p)$ and $\hat{\sigma}^*_{ij} = \hat{\sigma}_{ij} - \delta_{ij}/3$ are normalised quantities, δ_{ij} is the Kronecker delta and $\dot{\varepsilon}_v = \dot{\varepsilon}_{11} + \dot{\varepsilon}_{22} + \dot{\varepsilon}_{33}$ the volume strain rate. Function $\hat{a}$ in (4) is related to critical stress states which can be reached asymptotically under large shearing. With respect to the limit condition by Matsuoka and Nakai (1977) function $\hat{a}$ can be represented as (Bauer 2000):

$$\hat{a} = \frac{\sin\varphi_c}{3+\sin\varphi_c}\left[\sqrt{\frac{8/3 - 3(\hat{\sigma}^*_{kl}\hat{\sigma}^*_{kl}) + g\sqrt{3/2}(\hat{\sigma}^*_{kl}\hat{\sigma}^*_{kl})^{3/2}}{1 + g\sqrt{3/2}(\hat{\sigma}^*_{kl}\hat{\sigma}^*_{kl})^{1/2}}} + \sqrt{\hat{\sigma}^*_{kl}\hat{\sigma}^*_{kl}}\right] \tag{6}$$

with $g = -\sqrt{6}\hat{\sigma}^*_{kl}\hat{\sigma}^*_{lm}\hat{\sigma}^*_{mk}/(\hat{\sigma}^*_{pq}\hat{\sigma}^*_{pq})^{3/2}$. Herein φ_c denotes the intergranular friction angle defined for the critical state under triaxial compression. The influence of the mean effective pressure and the current void ratio on the response of the constitutive equation (4) is taken into account with the stiffness factor f_s and the density factor f_d. The dilatancy behavior, the peak stress ratio and strain softening depends on the density factor f_d which represents a relation between the current void ratio e, the critical void ratio e_c and the minimum void ratio e_d, i.e.

$$f_d = \left(\frac{e - e_d}{e_c - e_d}\right)^{\alpha} \, , \tag{7}$$

where $\alpha < 0.5$ is a constitutive constant. The stiffness factor f_s is proportional to the moisture dependent granular hardness $h^*_s = h_{so}\psi(w)$ and it also depends on mean effective stress p, i.e.

$$f_s = \left(\frac{e_i}{e}\right)^{\beta} \frac{h^*_s(1+e_i)}{nh_i(\hat{\sigma}_{kl}\hat{\sigma}_{kl})e_i}\left(\frac{3p}{h^*_s}\right)^{1-n} \, , \tag{8}$$

with

$$h_i = \frac{8\sin^2\varphi}{(3+\sin\varphi)^2} + 1 - \frac{2\sqrt{2}\sin\varphi}{3+\sin\varphi}\left(\frac{e_{io} - e_{do}}{e_{co} - e_{do}}\right)^{\alpha} \, .$$

Herein $\beta > 1$ is a constitutive constant. In (7) and (8) the current void ratio e is related to the pressure and moisture dependent maximum void ratio e_i, minimum void ratio e_d and critical void ratio e_c according to relation (3). A specific representation of the scalar function $\psi(w)$ can be obtained by curve fitting experimental data.

The proposed hypoplastic model for weathered broken rock materials includes nine constants which can be determined from simple index and element tests (Herle and Gudehus 1999, Bauer 1996). In the present paper the calibration of the constants is based on the experiments carried out by Kast (1992) with weathered broken granite. The following values were obtained: $\varphi_c = 42°$, $h_{so} = 75\,\text{MPa}$, $\psi = 1$ for dry states and $\psi = 0.34$ for the water saturated

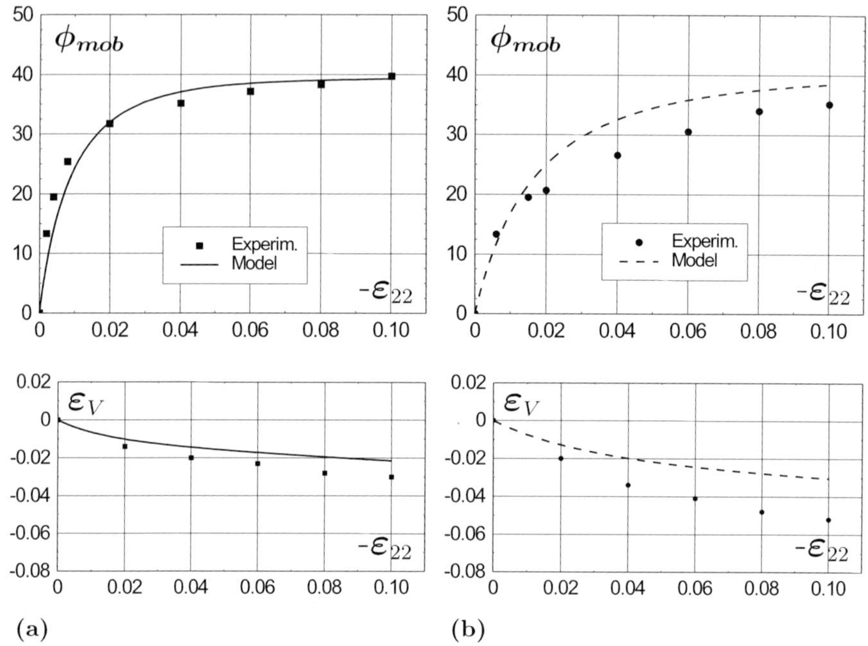

Fig. 3. Triaxial compression under a constant mean pressure starting from an initial void ratio of: (**a**) $e_0 = 0.418$ for the dry state ($\psi = 1$), (**b**) $e_0 = 0.387$ for the saturated state ($\psi = 0.34$)

state of the grains, $n = 0.6$, $e_{io} = 0.85$, $e_{co} = 0.39$, $e_{do} = 0.2$, $\alpha = 0.125$ and $\beta = 1.05$. It should be noted that in the present model the critical friction angle φ_c is assumed to be a constant because the experiments used for the present calibration did not show a clear influence of the moisture content on the critical friction angle.

3.1 Comparison of Numerical Simulations with Experiments

The performance of the proposed model is demonstrated by comparing the results obtained from the numerical simulation of homogeneous triaxial compression tests with the experiments carried out by Kast (1992). The so-called homogeneous element tests are performed for different precompactions of the material, i.e. different initial void ratios, and for dry and saturated states. In this context water saturation means the water saturation of the weathered grains and also the water saturation of the void space between the grains. In accordance with the experiments all simulations are performed under drained conditions. Neglecting the influence of the specific weights of the solid grains and of the water the effective grain stress are assumed to be equal to the total stress prescribed at the boundary of the specimen. Figures 3 and 4 show the numerical results obtained from the simulation of triaxial compression tests under a constant mean pressure of $p = 0.8\,\mathrm{MPa}$ and the experimental results obtained for weathered broken granite.

In particular for the loose material an initially void ratio of $e_0 = 0.418$ for the dry state and $e_0 = 0.387$ for the saturated state is considered. As the initial void ratio is higher than the corresponding pressure dependent critical one, i.e. for $p = 0.8\,\mathrm{MPa} \rightarrow e_c = 0.34$ for the dry material and $e_c = 0.31$ for the saturated material, the triaxial compression leads to a further densification for both the dry material (Fig. 3(**a**)) and the saturated material (Fig. 3(**b**)). The increase of the mobilized friction angle ϕ_{mob} with the vertical strain ε_{22} is more pronounced for the dry material than for the saturated one, which is also in agreement with the experiments. In order to study the influence of an initially dense material the experiments for triaxial compression under a constant mean pressure of $p = 0.8\,\mathrm{MPa}$ starting from $e = 0.29$ for the dry material and $e = 0.285$ for the saturated solid material are compared with the prediction of the hypoplastic model in Figs. 4(**a**) and (**b**). A comparison of Fig. 3(**a**) with Fig. 4(**a**) shows that the volume-strain behavior is strongly influenced by the initial density and it differs for the dry and saturated states of the solid material. The additional densification is less pronounced and the maximum mobilized friction angle is higher for the initially dense material.

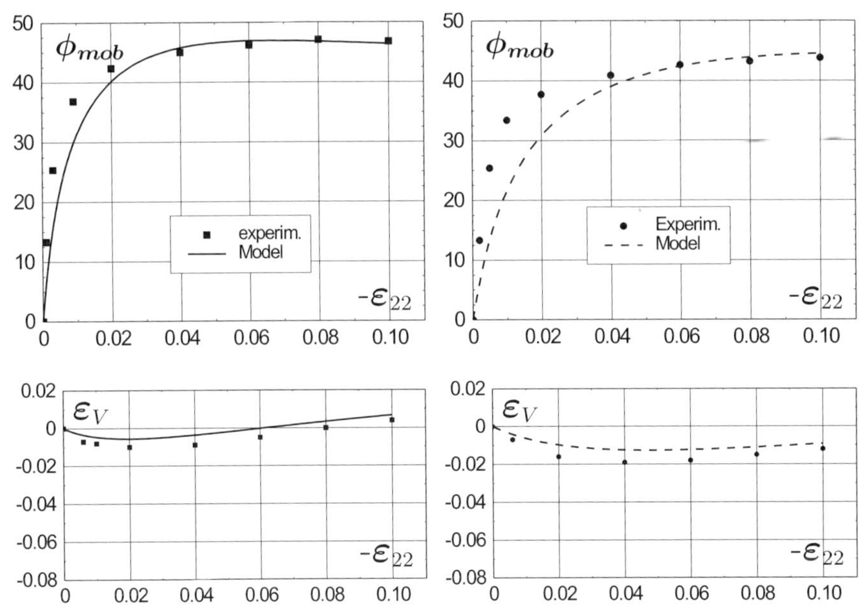

Fig. 4. Triaxial compression under a constant mean pressure starting from an initial void ratio of: (**a**) $e_0 = 0.29$ for the dry state ($\psi = 1$), (**b**) $e_0 = 0.285$ for the saturated state ($\psi = 0.34$)

A clear peak state for ϕ_{mob} can only be detected for the dry and initially dense material (Fig. 4(**a**)). After the peak the value of ϕ_{mob} slightly decreases with advanced vertical compression and it is accompanied by dilatancy. For the saturated and initially dense state of the material the dilatancy is less pronounced.

4 Conclusions

A hypoplastic continuum model is presented to describe the essential properties of weathered broken rock materials. In particular an increase of the compressibility and a decrease of the limit void ratios with an increase of the moisture content of the solid material is modeled in a simplified manner using only a moisture dependent granular hardness. It is demonstrated that the mechanical behavior of an initially loose and an initially dense material can be modeled with a single set of constants. The comparison of the numerical simulations of triaxial compression tests with experiments shows that the model captures the essential properties of weathered rockfill materials for both dry and water saturated grains.

References

Darve F (1991) Incrementally non-linear constitutive relationships. In: Darve (ed) Geomaterials: Constitutive Equations and Modelling. Elsevier:213–237.

Kolymbas D (1991) Archive of Applied Mechanics 3:143–151

Wu W, Bauer E (1993) A hypoplastic model for barotropy and pyknotropy of granular soils. In: Kolymbas D (ed) Proc Int Workshop on Modern Approaches to Plasticity. Elsevier:225–245

Wu W, Bauer E, Kolymbas D (1996) Mechanics of Materials 23:45–69

Herle I, Gudehus G (1999) Mechanics of Cohesive-Frictional Materials 4:461–486

Bauer E (2000) Mechanics of Cohesive-Frictional Materials 5:125–148

Gudehus G (1996) Soils and Foundations 36(1):1–12

Bauer E (1996) Soils and Foundations 36(1):13–26

Kast K (1992) Mechanisches Verhalten von Granitschüttungen. Heft 125: Veröffentlichungen des Institutes für Bodenmechanik und Felsmechanik der Universität Fridericiana in Karlsruhe, Germany

Bauer E, Zhu Y (2004) Constitutive modeling of the influence of pressure, density and moisture content on the mechanical behavior of rockfill materials. In: Wieland, Rena, Tan (eds) Proc 4th Int Conf on Dam Engineering, Nanjing Balkema:129–146.

Gudehus G (1997) Attractors, percolation thresholds and phase limits of granular soils. In: Behringer, Jenkins (eds) Powders and grains. Balkema:169–183

Matsuoka H, Nakai T (1977) Stress-strain relationship of soil based on the 'SMP'. In: Proc of Speciality Session 9, IX Int Conf Soil Mech Found Eng. Tokyo:153–162

A Framework for Analysis of Diffuse Instability in Partially Saturated Granular Soils

Ronaldo I. Borja

Department of Civil and Environmental Engineering, Stanford University, Stanford, CA 94305, USA borja@stanford.edu

Summary. Strain localization in the form of a deformation band is the most extensively investigated instability mode in geomaterials. Deformation is localized to a narrow band that could shear as well as either dilate or compact. The jump in the strain rate across the band is given by a tensor of determinant rank one. However, loose soils can also implode when subjected to compressive stresses. In this case the mechanism of deformation is diffuse in the sense that the jump in the strain rate tensor has a full rank. Only very recently has this mode been studied extensively in the context of material instability (Borja 2006a). In this paper we move one step further and consider the effect of volume constraint associated with the presence of fluids, specifically water and air, on loose soils that have a tendency to implode. The upshot is that, depending on the degree of saturation of these soils, the fluids may inhibit the solid from imploding freely resulting in a rise in the pore fluid pressures. The eigenmode at bifurcation is represented by the composite jumps in the strain rate tensor and the rates of pore air and pore water pressures. The framework of stability analysis presented in this work can be used to develop a criterion for the onset of liquefaction instability in fully saturated soils.

Key words: bifurcation, deformation bands, diffuse instability, partially saturated soils, porous media

1 Introduction

Strain localization is a ubiquitous feature of granular material behavior. In soils, zones of intense strain, called deformation bands, have been extensively investigated experimentally, theoretically, and numerically. This mode of instability is characterized by a jump in the velocity gradient of the form (Borja 2002)

$$[\![\boldsymbol{l}]\!] = \varphi \boldsymbol{m} \otimes \boldsymbol{n} \,, \tag{1}$$

where φ is a real multiplier representing the magnitude of velocity jump, and $\boldsymbol{m}$ and $\boldsymbol{n}$ are unit Cartesian vectors defining the instantaneous direction of velocity jump across the band and the direction of the normal to the band, respectively. In this case, $[\![\boldsymbol{l}]\!]$ is a tensor of determinant rank one, and its form may be determined from the associated singular tangent acoustic tensor.

On the other hand, equation (1) does not represent all possible instability modes. The jump tensor $[\![\boldsymbol{l}]\!]$ could have a full rank, equal to three for 3D problems, for example, implying that there could be three possible jumps in the principal stretch rates at the onset of instability (Borja 2006a). As an example, dry loose sands could implode in three directions, without manifesting a deformation band. Here, we have used the term 'implosion' and not 'compaction' since the former suggests some form of instability whereas the latter could merely be a simple reduction in volume due to compression. In the presence of fluids, soils undergoing volume implosion may have their pores inhibited from collapsing freely, resulting in jumps in the pore fluid pressures.

In this paper, we view volume implosion as an instability problem similar in spirit to a deformation band-type instability but allowing for the development of a full rank jump tensor characterizing the eigenmode, or e-mode, at bifurcation. The form of the full rank jump tensor $[\![\boldsymbol{l}]\!]$ is determined from the associated singular 'undrained' tangent constitutive tensor, and, formally, we classify such instability mode as diffuse. We formulate the problem by assuming that the pores of the soil matrix are filled with water and air. Depending on the degree of saturation of the solid-water-air mixture, the volume could implode freely or the mode of deformation could be subjected to a volume constraint imposed by the properties of the fluids trapped in the pores. As a limiting condition, the formulation presented in this paper could be used to define a criterion for the onset of liquefaction instability in fully saturated granular soils.

2 Balance of Mass

We refer to Borja (2006b, 2004) for details of the following formulation. Consider a solid matrix whose voids are continuous and filled with water and air. We denote the volume fraction by $\phi^\alpha = V_\alpha/V$ for $\alpha =$ solid, water and air (s, w and a respectively), where V_α is the portion of the total volume V of the mixture occupied by constituent α. The volume fractions satisfy the closure condition

$$\phi^{\mathrm{s}} + \phi^{\mathrm{w}} + \phi^{\mathrm{a}} = 1\,. \tag{2}$$

The partial mass density is $\rho^\alpha = \phi^\alpha \rho_\alpha$, where ρ_α is the intrinsic mass density, so the total mass density of the mixture is

$$\rho = \rho^{\mathrm{s}} + \rho^{\mathrm{w}} + \rho^{\mathrm{a}}\,. \tag{3}$$

We denote the time derivative with respect to the solid motion by

$$\frac{\mathrm{d}(\cdot)}{\mathrm{d}t} = \frac{\partial(\cdot)}{\partial t} + \mathrm{grad}(\cdot)\cdot\boldsymbol{v}\,, \tag{4}$$

where $\boldsymbol{v}$ is the velocity of the solid and grad is the spatial gradient operator. In this paper we shall use interchangeably the more abbreviated superposed dot and the time derivative with respect to the solid motion, i.e., $\dot{a} \equiv \mathrm{d}a/\mathrm{d}t$. The time derivative with respect to the fluid motion is given by

$$\frac{\mathrm{d}^\alpha(\cdot)}{\mathrm{d}t} = \frac{\partial(\cdot)}{\partial t} + \mathrm{grad}(\cdot) \cdot \boldsymbol{v}_\alpha \,, \qquad \alpha = \mathrm{w}, \mathrm{a} \,, \tag{5}$$

where $\boldsymbol{v}_\alpha$ is the velocity of fluid α. The time derivatives (4) and (5) are related by the equation

$$\frac{\mathrm{d}^\alpha(\cdot)}{\mathrm{d}t} = \frac{\mathrm{d}(\cdot)}{\mathrm{d}t} + \mathrm{grad}(\cdot) \cdot \widetilde{\boldsymbol{v}}_\alpha \,, \quad \widetilde{\boldsymbol{v}}_\alpha = \boldsymbol{v}_\alpha - \boldsymbol{v} \,. \tag{6}$$

In terms of the motion of the solid, balance of mass for the solid and fluid are given, respectively, by

$$\frac{\mathrm{d}\rho^{\mathrm{s}}}{\mathrm{d}t} + \rho^{\mathrm{s}} \,\mathrm{div}(\boldsymbol{v}) = 0 \,, \tag{7}$$

$$\frac{\mathrm{d}\rho^\alpha}{\mathrm{d}t} + \rho^\alpha \,\mathrm{div}(\boldsymbol{v}) = -\,\mathrm{div}(\boldsymbol{q}^\alpha) \,, \qquad \alpha = \mathrm{w}, \mathrm{a} \,, \tag{8}$$

where $\boldsymbol{q}^\alpha = \rho^\alpha \widetilde{\boldsymbol{v}}_\alpha$ is the Eulerian relative flow vector of fluid α relative to the solid, and div is the spatial divergence operator.

For barotropic flows the intrinsic bulk modulus of the constituent α can be defined as

$$K_\alpha = \rho_\alpha p'_\alpha(\rho_\alpha) \,, \qquad \alpha = \mathrm{s}, \mathrm{w}, \mathrm{a} \,, \tag{9}$$

where p_α is the intrinsic Cauchy pressure in the α constituent (compressive normal force acting on this constituent per unit area of the same constituent). Here we assume that p_α has a functional relationship with the intrinsic mass density ρ_α, and the bulk modulus K_α is an intrinsic property of the material. Inserting (9) into (7) and (8) gives

$$\frac{\mathrm{d}\phi^{\mathrm{s}}}{\mathrm{d}t} + \frac{\phi^{\mathrm{s}}}{K_{\mathrm{s}}} \frac{\mathrm{d}p_{\mathrm{s}}}{\mathrm{d}t} + \phi^{\mathrm{s}} \,\mathrm{div}(\boldsymbol{v}) = 0 \,, \tag{10}$$

$$\frac{\mathrm{d}\phi^\alpha}{\mathrm{d}t} + \frac{\phi^\alpha}{K_\alpha} \frac{\mathrm{d}p_\alpha}{\mathrm{d}t} + \phi^\alpha \,\mathrm{div}(\boldsymbol{v}) = -\frac{1}{\rho_\alpha} \mathrm{div}(\boldsymbol{q}^\alpha) \,, \qquad \alpha = \mathrm{w}, \mathrm{a} \,. \tag{11}$$

Next we introduce a bulk modulus K representing the volumetric stiffness of the soil matrix, or skeleton. This quantity must be distinguished from the bulk modulus K_{s} introduced earlier which represents the volumetric stiffness of the solid grains themselves. Reference Borja (2006b) presents a closed-form expression for K from an assumed functional relationship among the intrinsic variables p_{s} and ρ_{s}, and the solid volume fraction ϕ^{s}. The bulk modulus K is defined such that

$$\phi^{\mathrm{s}} \frac{\mathrm{d}p_{\mathrm{s}}}{\mathrm{d}t} = -K \,\mathrm{div}(\boldsymbol{v}) \,. \tag{12}$$

Inserting (12) into (10) yields the following alternative form of balance of mass for the solid

$$\frac{\mathrm{d}\phi^{\mathrm{s}}}{\mathrm{d}t} + \left(\phi^{\mathrm{s}} - \frac{K}{K_{\mathrm{s}}}\right) \mathrm{div}(\boldsymbol{v}) = 0\,. \tag{13}$$

Adding (13) to (11) yields the balance of mass for the entire solid-water-air mixture,

$$\sum_{\alpha=\mathrm{w,a}} \frac{\phi^\alpha}{K_\alpha}\frac{\mathrm{d}p_\alpha}{\mathrm{d}t} + B\,\mathrm{div}(\boldsymbol{v}) = -\sum_{\alpha=\mathrm{w,a}} \frac{1}{\rho_\alpha}\,\mathrm{div}(\boldsymbol{q}^\alpha)\,, \tag{14}$$

where

$$B := 1 - \frac{K}{K_{\mathrm{s}}} \tag{15}$$

is the Biot coefficient. For soils $K \ll K_{\mathrm{s}}$, and B may be assumed equal to unity. However, we shall continue to use the parameter B to preserve a more general form of the governing equations.

Next we introduce void fractions ψ^{w} and ψ^{a}, defined as the respective ratios between the volumes of the water and air phases in the void to the volume of the void itself,

$$\psi^{\mathrm{w}} = \frac{\phi^{\mathrm{w}}}{1-\phi^{\mathrm{s}}}\,, \qquad \psi^{\mathrm{a}} = \frac{\phi^{\mathrm{a}}}{1-\phi^{\mathrm{s}}}\,. \tag{16}$$

The void fraction ψ^{w} is often called the degree of saturation, commonly denoted by the symbol S_{r} in the literature. The saturation condition is

$$\psi^{\mathrm{w}} + \psi^{\mathrm{a}} = 1\,. \tag{17}$$

Taking the material time derivative of (16) in the direction of the solid motion and using (13) gives

$$\begin{aligned} \frac{\mathrm{d}\phi^\alpha}{\mathrm{d}t} &= (1-\phi^{\mathrm{s}})\frac{\mathrm{d}\psi^\alpha}{\mathrm{d}t} - \psi^\alpha\frac{\mathrm{d}\phi^{\mathrm{s}}}{\mathrm{d}t} \\ &= (1-\phi^{\mathrm{s}})\frac{\mathrm{d}\psi^\alpha}{\mathrm{d}t} + \psi^\alpha\left(\phi^{\mathrm{s}} - \frac{K}{K_{\mathrm{s}}}\right)\mathrm{div}(\boldsymbol{v}) \end{aligned} \tag{18}$$

for $\alpha = \mathrm{w,a}$. Substituting this result into (11) yields

$$(1-\phi^{\mathrm{s}})\frac{\mathrm{d}\psi^\alpha}{\mathrm{d}t} + \frac{\phi^\alpha}{K_\alpha}\frac{\mathrm{d}p_\alpha}{\mathrm{d}t} + \psi^\alpha B\,\mathrm{div}(\boldsymbol{v}) = -\frac{1}{\rho_\alpha}\,\mathrm{div}(\boldsymbol{q}^\alpha)\,. \tag{19}$$

For finite deformation it is usually convenient to transform the pressure variables. To this end, we let $\boldsymbol{F}$ denote the deformation gradient of the solid motion and $J = \det(\boldsymbol{F})$ the corresponding Jacobian. We define the Kirchhoff pore fluid pressure as

$$\theta_\alpha = Jp_\alpha\,, \qquad \alpha = \mathrm{w,a}\,, \tag{20}$$

with material time derivative, relative to the solid motion, given by

$$\dot{\theta}_\alpha \equiv \frac{\mathrm{d}\theta_\alpha}{\mathrm{d}t} = J\dot{p}_\alpha + \dot{J}p_\alpha \,. \tag{21}$$

Hence, multiplying both sides of (19) by J and using the Piola identity gives

$$J(1-\phi^{\mathrm{s}})\dot{\psi}^\alpha + \frac{\phi^\alpha}{K_\alpha}\dot{\theta}_\alpha + \left(\psi^\alpha B - \frac{\phi^\alpha}{K_\alpha}p_\alpha\right)\dot{J} = -\frac{1}{\rho_\alpha}\,\mathrm{DIV}(\boldsymbol{Q}^\alpha)\,, \tag{22}$$

where

$$\boldsymbol{Q}^\alpha = J\boldsymbol{F}^{-1}\cdot\boldsymbol{q}^\alpha\,, \qquad \alpha = \mathrm{w},\mathrm{a} \tag{23}$$

is the Piola transform of $\boldsymbol{q}^\alpha$ and DIV is the divergence operator evaluated with respect to the reference configuration.

Next we consider a constitutive law describing the degree of saturation versus capillary pressure (or suction stress) relation. Taking $\psi^{\mathrm{w}} = S_{\mathrm{r}}$ and $\psi^{\mathrm{a}} = 1 - S_{\mathrm{r}}$, the constitutive law takes the form

$$S_{\mathrm{r}} = S_{\mathrm{r}}(s)\,, \qquad s = \theta_{\mathrm{a}} - \theta_{\mathrm{w}}\,. \tag{24}$$

In the above equation, s is the Kirchhoff suction stress, or Kirchhoff capillary pressure. Taking the time derivative with respect to the solid motion gives

$$\dot{\psi}^{\mathrm{w}} = -\dot{\psi}^{\mathrm{a}} = \dot{S}_{\mathrm{r}} = S_{\mathrm{r}}'(s)(\dot{\theta}_{\mathrm{a}} - \dot{\theta}_{\mathrm{w}})\,. \tag{25}$$

Inserting this result into (22) gives the pair of equations

$$a_{11}\dot{\theta}_{\mathrm{w}} + a_{12}\dot{\theta}_{\mathrm{a}} + b_1\dot{J} = -\frac{1}{\rho_{\mathrm{w}}}\,\mathrm{DIV}(\boldsymbol{Q}^{\mathrm{w}})\,, \tag{26}$$

$$a_{12}\dot{\theta}_{\mathrm{w}} + a_{22}\dot{\theta}_{\mathrm{a}} + b_2\dot{J} = -\frac{1}{\rho_{\mathrm{a}}}\,\mathrm{DIV}(\boldsymbol{Q}^{\mathrm{a}})\,. \tag{27}$$

where

$$\begin{gathered} a_{11} = \frac{\phi^{\mathrm{w}}}{K_{\mathrm{w}}} - a_{12}\,, \qquad a_{22} = \frac{\phi^{\mathrm{a}}}{K_{\mathrm{a}}} - a_{12}\,, \qquad a_{12} = J(1-\phi^{\mathrm{s}})S_{\mathrm{r}}'(s)\,, \\ b_1 = \psi^{\mathrm{w}}B - \frac{\phi^{\mathrm{w}}}{K_{\mathrm{w}}}p_{\mathrm{w}}, \qquad b_2 = \psi^{\mathrm{a}}B - \frac{\phi^{\mathrm{a}}}{K_{\mathrm{a}}}p_{\mathrm{a}}\,. \end{gathered} \tag{28}$$

In the next section we shall provide a physical significance of the Piola transforms $\boldsymbol{Q}^{\mathrm{w}}$ and $\boldsymbol{Q}^{\mathrm{a}}$.

3 Balance of Linear Momentum

Without loss of generality we shall consider a quasi-static loading condition and write the balance of linear momentum in Lagrangian form as

$$\mathrm{DIV}(\boldsymbol{P}) + \rho_0\mathbf{g} = \mathbf{0}\,, \qquad \rho_0 = J\rho\,, \tag{29}$$

where $\boldsymbol{P}$ is the first Piola–Kirchhoff total stress tensor, $\mathbf{g}$ is the gravity acceleration vector, and ρ is the saturated mass density of the mixture (cf. (3)). For bifurcation analysis it is more useful to consider the rate form of (29), which is given by

$$\mathrm{DIV}(\dot{\boldsymbol{P}}) + \dot{\rho}_0 \mathbf{g} = \mathbf{0}\,, \tag{30}$$

where

$$\begin{aligned} \dot{\rho}_0 \equiv \frac{\mathrm{d}(J\rho)}{\mathrm{d}t} &= J \sum_{\alpha=\mathrm{s,w,a}} \dot{\rho}^\alpha + \dot{J} \sum_{\alpha=\mathrm{s,w,a}} \rho^\alpha \\ &= \sum_{\alpha=\mathrm{s,w,a}} J\Big(\dot{\rho}^\alpha + \rho^\alpha\,\mathrm{DIV}(\boldsymbol{v})\Big) \\ &= -\sum_{\alpha=\mathrm{w,a}} J\,\mathrm{DIV}(\boldsymbol{q}^\alpha) = -\sum_{\alpha=\mathrm{w,a}} \mathrm{DIV}(\boldsymbol{Q}^\alpha)\,, \end{aligned} \tag{31}$$

after using the mass balance equations (7) and (8). Hence, the sum of the Piola transforms $\boldsymbol{Q}^\alpha$ for $\alpha = \mathrm{w, a}$ has the physical significance that its divergence with respect to the reference configuration is the negative of the time derivative (with respect to the solid motion) of the pull-back mass density ρ_0. Note that for a mixture of two or more constituents, $\dot{\rho}_0 = 0$ if and only if there is no relative flow of the fluid relative to the solid matrix.

A critical aspect of balance of linear momentum for a mixture of two or more constituents lies in the decomposition of the total stress tensor. To this end we recall the following stress tensor decomposition emerging from continuum principles of thermodynamics in the absence of non-mechanical energy for a solid matrix whose voids are filled with water and air (Borja 2006b):

$$\boldsymbol{\sigma} = \boldsymbol{\sigma}' - B\overline{p}\mathbf{1}\,, \tag{32}$$

where $\boldsymbol{\sigma}$ is the total Cauchy stress tensor, $\boldsymbol{\sigma}'$ is a constitutive (or effective) Cauchy stress that is energy-conjugate to the rate of deformation of the solid matrix, $\mathbf{1}$ is the second order identity tensor (Kronecker delta), and

$$\overline{p} = \sum_{\alpha=\mathrm{w,a}} \psi^\alpha p_\alpha = S_\mathrm{r} p_\mathrm{w} + (1 - S_\mathrm{r}) p_\mathrm{a} \tag{33}$$

is the mean Cauchy pore pressure obtained from the pore air and pore water pressures weighted according to the degree of saturation. As a matter of sign convention, a positive normal stress implies tension in the present case.

Multiplying both sides of (32) by J gives

$$\boldsymbol{\tau} = \boldsymbol{\tau}' - B\overline{\theta}\mathbf{1}\,, \tag{34}$$

where $\boldsymbol{\tau} = J\boldsymbol{\sigma}$ and $\boldsymbol{\tau}' = J\boldsymbol{\sigma}'$ are the symmetric total and effective Kirchhoff stress tensors, respectively, and

$$\overline{\theta} = J\overline{p} = S_\mathrm{r}\theta_\mathrm{w} + (1 - S_\mathrm{r})\theta_\mathrm{a}\,. \tag{35}$$

Contracting the right index of $\boldsymbol{\tau}$ by $\boldsymbol{F}^{-\mathrm{t}}$ gives

$$\boldsymbol{P} = \boldsymbol{P}' - B\overline{\theta}\boldsymbol{F}^{-\mathrm{t}}\,, \tag{36}$$

where $\boldsymbol{P} = \boldsymbol{\tau} \cdot \boldsymbol{F}^{-\mathrm{t}}$ and $\boldsymbol{P}' = \boldsymbol{\tau}' \cdot \boldsymbol{F}^{-\mathrm{t}}$. Taking the time derivative of (36) with respect to the solid motion gives

$$\dot{\boldsymbol{P}} = \dot{\boldsymbol{P}}' - B\overline{\theta}\dot{\boldsymbol{F}}^{-\mathrm{t}} - B\dot{\overline{\theta}}\boldsymbol{F}^{-\mathrm{t}} - \dot{B}\overline{\theta}\boldsymbol{F}^{-\mathrm{t}}\,, \tag{37}$$

where $\dot{B} = -\dot{K}/K_{\mathrm{s}}$ (assuming K_{s} is constant), and

$$\dot{\overline{\theta}} = \overline{S}_r\dot{\theta}_{\mathrm{w}} + (1 - \overline{S}_{\mathrm{r}})\dot{\theta}_{\mathrm{a}}\,, \qquad \overline{S}_{\mathrm{r}} = S_{\mathrm{r}} + S_{\mathrm{r}}'(s)s\,. \tag{38}$$

We now put these results together. Let the initial placement of a partially saturated solid body be denoted by $\mathcal{B}$, and let $\partial\mathcal{B}$ denote its boundary such that $\partial\mathcal{B} = \overline{\partial\mathcal{B}_t \cup \partial\mathcal{B}_u}$ and $\emptyset = \partial\mathcal{B}_t \cap \partial\mathcal{B}_u$, where $\partial\mathcal{B}_t$ and $\partial\mathcal{B}_u$ are portions of the entire boundary where nominal traction rates and velocities are prescribed. We denote the solution pair by $(\dot{\boldsymbol{P}}, \boldsymbol{v})$. For this solution to be admissible we must have

$$\mathrm{DIV}(\dot{\boldsymbol{P}}) - \sum_{\alpha=\mathrm{w,a}} \mathrm{DIV}(\boldsymbol{Q}^\alpha)\mathbf{g} = \mathbf{0} \qquad \text{in} \quad \mathcal{B} \tag{39}$$

$$\boldsymbol{v} = \dot{\boldsymbol{u}}_0 \quad \text{on} \quad \partial\mathcal{B}_u \tag{40}$$

$$\dot{\boldsymbol{P}} \cdot \boldsymbol{N} = \dot{\boldsymbol{t}}_0 \quad \text{on} \quad \partial\mathcal{B}_t \tag{41}$$

where $\boldsymbol{N}$ is the unit outward normal vector to $\partial\mathcal{B}$, and $\boldsymbol{u}_0$ and $\boldsymbol{t}_0$ are prescribed solid boundary displacements and total nominal tractions, respectively. Note that the partial differential equation and boundary conditions are written in terms of the total (and not effective) stress rate tensor $\dot{\boldsymbol{P}}$.

4 Local Stability and Uniqueness

Consider an alternative solution pair $(\dot{\boldsymbol{P}}^*, \boldsymbol{v}^*) \neq (\dot{\boldsymbol{P}}, \boldsymbol{v})$ satisfying the partial differential equation (39) and boundary conditions (40) and (41). To investigate the existence of this alternative solution, we consider the integral

$$\begin{aligned}\int_{\mathcal{B}} \mathrm{DIV}[(\boldsymbol{v}^* - \boldsymbol{v}) \cdot (\dot{\boldsymbol{P}}^* - \dot{\boldsymbol{P}})]\,\mathrm{d}V &= \int_{\mathcal{B}} (\dot{\boldsymbol{F}}^* - \dot{\boldsymbol{F}}) : (\dot{\boldsymbol{P}}^* - \dot{\boldsymbol{P}})\,\mathrm{d}V \\ &+ \int_{\mathcal{B}} (\boldsymbol{v}^* - \boldsymbol{v}) \cdot [\mathrm{DIV}(\dot{\boldsymbol{P}}^*) - \mathrm{DIV}(\dot{\boldsymbol{P}})]\,\mathrm{d}V\,.\end{aligned} \tag{42}$$

Using Gauss theorem the integral on the left-hand side of (42) is

$$\int_{\mathcal{B}} \mathrm{DIV}[(\boldsymbol{v}^* - \boldsymbol{v}) \cdot (\dot{\boldsymbol{P}}^* - \dot{\boldsymbol{P}})]\,\mathrm{d}V = \int_{\partial\mathcal{B}} (\boldsymbol{v}^* - \boldsymbol{v}) \cdot (\dot{\boldsymbol{P}}^* - \dot{\boldsymbol{P}}) \cdot \boldsymbol{N}\,\mathrm{d}A = 0\,, \tag{43}$$

since both solutions are required to satisfy the boundary conditions (40) and (41). The second integral on the right-hand side is zero since both solutions are required to satisfy the partial differential equation (39) for fixed flow vectors $\boldsymbol{Q}^\alpha$.

That the flow vectors $\boldsymbol{Q}^\alpha$ are the same irrespective of whether one chooses the solid velocity $\boldsymbol{v}$ or $\boldsymbol{v}^*$ implies that a jump in the solid velocity is not accompanied by a jump in the fluid flow. In short, we are interested in 'undrained' bifurcation. In this case, any pair of possible solutions must satisfy the condition

$$\int_{\mathcal{B}} (\dot{\boldsymbol{F}}^* - \dot{\boldsymbol{F}}) : (\dot{\boldsymbol{P}}^* - \dot{\boldsymbol{P}})\, \mathrm{d}V = 0\,. \tag{44}$$

Uniqueness is guaranteed for every point and every pair of stresses and deformation gradients linked by the constitutive equation if

$$(\dot{\boldsymbol{F}}^* - \dot{\boldsymbol{F}}) : (\dot{\boldsymbol{P}}^* - \dot{\boldsymbol{P}}) > 0\,. \tag{45}$$

If this is the case, then the material is incrementally stable.

Uniqueness is lost when the jump in the first Piola–Kirchhoff stress rate vanishes,

$$[\![\dot{\boldsymbol{P}}]\!] = \dot{\boldsymbol{P}}^* - \dot{\boldsymbol{P}} = \boldsymbol{0}\,. \tag{46}$$

For incrementally linear systems where the tangent constitutive tensor has the major symmetry the criterion of stationary stress rate coincides with initial loss of local incremental stability in the sense of definition (45). However, for nonsymmetric systems condition (45) could be violated well before condition (46), see Borja (2006a), for example. Since a non-stationary stress state violates the equilibrium condition, we take (46) as our relevant criterion for loss of incremental stability. An analog of this criterion is the choice of loss of ellipticity to detect a deformation band-type bifurcation since for nonsymmetric systems the loss of strong ellipticity does not guarantee a stationary traction rate, again see Borja (2006a).

The formulation simplifies considerably in the spatial description. Recall that $\boldsymbol{\tau} = \boldsymbol{P} \cdot \boldsymbol{F}^{\mathrm{t}}$; taking the time derivatives and evaluating the jumps in the rates we obtain

$$[\![\dot{\boldsymbol{\tau}}]\!] = \boldsymbol{P} \cdot [\![\dot{\boldsymbol{F}}^{\mathrm{t}}]\!] + [\![\dot{\boldsymbol{P}}]\!] \cdot \boldsymbol{F}^{\mathrm{t}}\,, \tag{47}$$

where

$$[\![\dot{\boldsymbol{F}}]\!] = [\![\boldsymbol{l}]\!] \cdot \boldsymbol{F}\,, \qquad [\![\boldsymbol{l}]\!] = \frac{\partial [\![\boldsymbol{v}]\!]}{\partial \boldsymbol{x}} = \frac{\partial (\boldsymbol{v}^* - \boldsymbol{v})}{\partial \boldsymbol{x}} = \boldsymbol{l} - \boldsymbol{l}^*\,, \tag{48}$$

and $[\![\boldsymbol{l}]\!]$ is the jump in the velocity gradient. Setting $[\![\dot{\boldsymbol{P}}]\!] = \boldsymbol{0}$ in (47) results in the equivalent condition for a stationary first Piola–Kirchhoff stress rate

$$[\![\dot{\boldsymbol{\tau}}]\!] - (\boldsymbol{\tau} \ominus \boldsymbol{1}) : [\![\boldsymbol{l}]\!] = \boldsymbol{0}\,, \tag{49}$$

where $(\boldsymbol{\tau} \ominus \boldsymbol{1})_{ijkl} = \tau_{il}\delta_{jk}$, and δ_{jk} = Kronecker delta. The jump in the total Kirchhoff stress rate in this case may be obtained from (34) as

$$[\![\dot{\boldsymbol{\tau}}]\!] = [\![\dot{\boldsymbol{\tau}}']\!] - [\![\dot{B}]\!]\bar{\theta}\mathbf{1} - B[\![\dot{\bar{\theta}}]\!]\mathbf{1}\,, \qquad [\![\dot{B}]\!] = -[\![\dot{K}]\!]/K_{\mathrm{s}}\,. \tag{50}$$

We now consider the following incremental constitutive equation

$$\dot{\boldsymbol{\tau}}' = \boldsymbol{\alpha} : \boldsymbol{l}\,, \qquad \dot{\boldsymbol{\tau}}'^{*} = \boldsymbol{\alpha} : \boldsymbol{l}^{*}\,, \tag{51}$$

where $\boldsymbol{\alpha}$ is a rank-four tensor with minor symmetry on its first two left indices but not on its last two right indices. We note two important points from the above relations: (a) the constitutive equation relates the Kirchhoff effective stress rate tensor $\dot{\boldsymbol{\tau}}'$ with the velocity gradient $\boldsymbol{l}$ as motivated by continuum principles of thermodynamics; and (b) the same tangent constitutive tensor $\boldsymbol{\alpha}$ is used irrespective of the direction of the velocity gradient. The latter point characterizes an incrementally linear material whose implication to elastoplasticity can be extended in the context of "in-loading comparison solid" (Borja 2006a). For completeness we shall also assume the following constitutive law for the bulk modulus K of the solid matrix (cf. (12)):

$$\dot{K} = C\phi^{\mathrm{s}}\dot{p}_{\mathrm{s}} = -CK\,\mathrm{DIV}(\boldsymbol{v}) = -CK\,\mathrm{tr}(\boldsymbol{l})\,, \tag{52}$$

where C is a positive coefficient. In words, this relation simply states that the increment of the bulk modulus of the solid matrix is linearly proportional to $\phi^{\mathrm{s}}\dot{p}_{\mathrm{s}}$ representing the increment of the intrinsic solid pressure "smeared" by the solid volume fraction.

Combining the above results yields the following condition for a stationary stress rate

$$\boldsymbol{\beta} : [\![\boldsymbol{l}]\!] - B[\![\dot{\bar{\theta}}]\!]\mathbf{1} = \mathbf{0}\,, \qquad \boldsymbol{\beta} := \boldsymbol{\alpha} - \boldsymbol{\tau} \ominus \mathbf{1} - C\theta\frac{K}{K_{\mathrm{s}}}\mathbf{1}\otimes\mathbf{1}\,. \tag{53}$$

Further, in order for the solution pair $(\dot{\boldsymbol{P}}^{*}, \boldsymbol{v}^{*})$ to be admissible it also must satisfy balance of mass, equations (26) and (27), herein written in jump form for fixed flow vectors $\boldsymbol{Q}^{\alpha}$ as

$$a_{11}[\![\dot{\theta}_{\mathrm{w}}]\!] + a_{12}[\![\dot{\theta}_{\mathrm{a}}]\!] + b_1[\![\dot{J}]\!] = 0\,, \tag{54}$$

$$a_{12}[\![\dot{\theta}_{\mathrm{w}}]\!] + a_{22}[\![\dot{\theta}_{\mathrm{a}}]\!] + b_2[\![\dot{J}]\!] = 0\,. \tag{55}$$

Solving for $[\![\dot{\theta}_{\mathrm{w}}]\!]$ and $[\![\dot{\theta}_{\mathrm{w}}]\!]$ yields

$$[\![\dot{\theta}_{\mathrm{w}}]\!] = c_{\mathrm{w}}[\![\dot{J}]\!]\,, \qquad [\![\dot{\theta}_{\mathrm{a}}]\!] = c_{\mathrm{a}}[\![\dot{J}]\!], \tag{56}$$

where

$$c_{\mathrm{w}} = \frac{a_{12}b_2 - a_{22}b_1}{a_{11}a_{22} - a_{12}^2}\,, \qquad c_{\mathrm{a}} = \frac{a_{12}b_1 - a_{11}b_2}{a_{11}a_{22} - a_{12}^2}\,. \tag{57}$$

Therefore, the jump in the mean pore pressure rate is

$$[\![\dot{\bar{\theta}}]\!] = \overline{S}_{\mathrm{r}}[\![\dot{\theta}_{\mathrm{w}}]\!] + (1 - \overline{S}_{\mathrm{r}})[\![\dot{\theta}_{\mathrm{a}}]\!] = -\lambda\mathbf{1} : [\![\boldsymbol{l}]\!]\,, \tag{58}$$

where

$$\lambda = -J[\,\overline{S}_{\mathrm{r}} c_{\mathrm{w}} + (1 - \overline{S}_{\mathrm{r}}) c_{\mathrm{a}}\,] > 0\,. \tag{59}$$

Finally, we substitute (58) into (53) to obtain the following alternative condition for a stationary stress rate,

$$(\boldsymbol{\beta} + B\lambda \mathbf{1} \otimes \mathbf{1}) : [\![\boldsymbol{l}]\!] = \mathbf{0}\,. \tag{60}$$

For a non-trivial solution $[\![\boldsymbol{l}]\!] \neq \mathbf{0}$ to exist we must then have

$$\det(\boldsymbol{\beta} + B\lambda \mathbf{1} \otimes \mathbf{1}) = 0\,. \tag{61}$$

Equation (61) is the condition for diffuse instability in partially saturated granular soils. This condition depends on the "drained"constitutive tangent material response as reflected by the value of $\boldsymbol{\beta}$, and on the mean bulk stiffness of the pores as represented by the overall bulk stiffness λ. In the limit of full saturation $S_{\mathrm{r}} = 1$, and condition (61) may be used to determine the onset of liquefaction instability in granular soils, see Borja (2006c).

5 Closure

We have presented a framework for mathematical analysis of a class of instability modes characterized by a full rank jump tensor in partially saturated granular soils. In the limit of zero saturation the kinematical signature is characterized by volume implosion in which jumps in the velocity gradient tensor occur in three principal directions. This kinematical signature is in stark contrast to the more extensively studied problem of deformation banding in which the jump in the velocity gradient tensor has a unit rank. Balance of mass for the fluids is used to provide a volume constraint to the kinematical jumps. In the limit of full saturation one may expect a nearly isochoric jump in the velocity gradient tensor accompanied by a jump in the pore water pressure due to the high bulk modulus of water, precisely the instability mode associated with the onset of liquefaction instability.

Acknowledgement

This work has been supported in part by the U.S. National Science Foundation, Grant Nos. CMS-0201317 and CMS-0324674.

References

Borja RI (2006a) Acta Geotechnica 1:107–122
Borja RI (2002) Comput Methods Appl Mech Engrg 191:5287–5314
Borja RI (2006b) Int J Solids Struc 43:1764–1768
Borja RI (2004) Comput Methods Appl Mech Engrg 193:5301–5338
Borja RI (2006c) Acta Geotechnica 1: in press

Interpretation of the Behaviour of Compacted Soils Using Cam-Clay Extended to Unsaturated Conditions

Francesca Casini[1], Roberto Vassallo[2], Claudio Mancuso[3], and Augusto Desideri[1]

[1] Dipartimento di Ingegneria Strutturale e Geotecnica, Università degli Studi di Roma "La Sapienza", Italy
francesca.casini@uniroma1.it, augusto.desideri@uniroma1.it

[2] Dipartimento di Strutture, Geotecnica, Geologia Applicata all'Ingegneria, Università degli Studi della Basilicata, Italy roberto.vassallo@unibas.it

[3] Dipartimento di Ingegneria Geotecnica, Università degli Studi di Napoli Federico II, Italy claudio.mancuso@unina.it

Summary. This paper is focused on the interpretation of experimental results obtained at the Department of Geotechnical Engineering of the University of Naples Federico II (Italy) to investigate the effects of partial saturation on the volumetric behaviour and the initial shear stiffness of a compacted silt. Tests were performed by using suction-controlled triaxial and resonant column cells. Herein, the compatibility of the results with a Single Stress Model (SSM) is discussed. The SSM allows to highlight that suction can have two effects on the mechanical behaviour of an unsaturated soil: it increases the average volumetric stress acting on the soil skeleton and it has a sort of cementing effect on the soil packing (hardening and cementation).

Key words: compacted soil, constitutive modelling, single stress variable, elasto-plastic, compressibility, small strain behaviour

1 Introduction

The experimental results obtained at the University of Naples Federico II (Italy) for the characterization of the small and moderate strain-behaviour of a compacted silt (Vassallo et al. 2006a) are interpreted in this paper using a Modified Cam Clay Model extended to unsaturated conditions (Jommi and di Prisco 1994, Tamagnini 2004).

The tested soil is representative of the materials used for constructing embankments on the Po river (Italy). It has a liquid limit $w_L = 50.4\%$ and a plasticity index $IP = 17.9\%$. The soil was compacted at the optimum water content with the standard Proctor procedure (ASTM D691–91). On average, $w = 23.12\%$ and $\gamma_d = 15.60\,\mathrm{kN/m^3}$ were obtained.

All tests included an equalization, during which suction changed from the after compaction value to the desired value, at mean net stress $p-u_a$ of about 10 kPa. Then, the imposed suction was kept constant during compression. Finally, in some cases, wetting and/or drying stages at constant $(p-u_a)$ were carried out (Vassallo et al. 2006a).

Experimental results were analysed by Vassallo et al. (2006b) using two independent stress variables and the framework of hardening plasticity.

2 Modelling Framework

The constitutive laws for the mechanical behaviour are defined in terms of effective stresses. The average soil skeleton stress p' is defined as the difference between total stress and an equivalent fluid pressure, with the degree of saturation as weighing parameter:

$$p' = p - u_a + S_r(u_a - u_w)\,. \tag{1}$$

The evolution of the scalar internal variable p'_c (overconsolidation pressure) depends not only on the rate of plastic strains but also on the variations of degree of saturation:

$$\dot{p}'_c = \dot{p}'_{\mathrm{csat}} - b' p_c \dot{S}_r \tag{2}$$

where b is a constant soil property. The integration of Eq. (2) yields to:

$$p'_c = p'_{\mathrm{csat}} \mathrm{e}^{b(1-S_r)}\,. \tag{3}$$

Thus, b controls the rate of change in p'_c caused by variations in S_r.

Hardening has an irreversible component dependent on the development of plastic volumetric strains, related to the evolution of p'_{csat}, and a reversible component related to changes in S_r.

The model predicts that a drying process induces some *bonding* and positive hardening while a wetting process induces some *debonding* (negative hardening), as shown in Fig. 1.

Let us consider an isotropic test in which a soil sample is subjected to an increment of p' at the virgin state. Similarly to saturated soils, the specific volume $v = 1 + e$ will be given by:

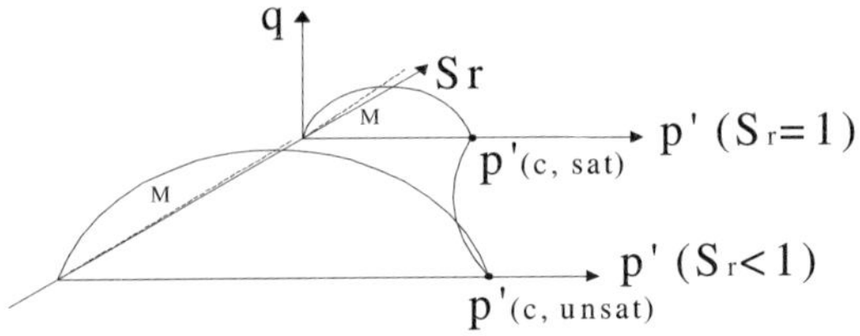

Fig. 1. Evolution of the hardening surface in the space $p' : S_r : q$

$$v = N(S_r) - \lambda \ln p' \tag{4}$$

where p' is the stress defined by Eq. (1) and $N(S_r)$ is the specific volume at $p' = 1\,\text{kPa}$.

The relationship which defines how N changes with the degree of saturation can be obtained from previous equations and from:

$$\begin{aligned} v_s &= N - \lambda \ln p'_{\text{csat}} \,, \\ v_s &= v - \kappa \ln \frac{p'_{\text{csat}}}{p'_c} \end{aligned} \tag{5}$$

where v_s is the specific volume of the saturated soil, N (equal to 2.01 for the considered soil) is the specific volume at $p' = 1\,\text{kPa}$ in saturated conditions. By substitution, one obtains:

$$N(S_r) = N + (\lambda - \kappa) b (1 - S_r) \tag{6}$$

If there is a variation of degree of saturation, this corresponds to a translation of the Normal Consolidation Line, parallel to the saturated NCL (Fig. 2). The amount of translation is related to the constitutive parameter b. By interpreting all the available compression stages (25, in total), an average value of 7 was obtained.

The model requires an hydraulic constitutive relationship describing the water storage mechanism, such as $S_r = WRC(ns)$ (Houlsby 1997, Tamagnini 2004), where ns is the product of soil porosity and suction. In this study, a modified form of the equation proposed by Van Genuchten (1980) is used:

$$S_r = \left[\frac{1}{1 + (\alpha n s)^r} \right]^t \tag{7}$$

whit $r = 1$, $t = 0.8$, and:

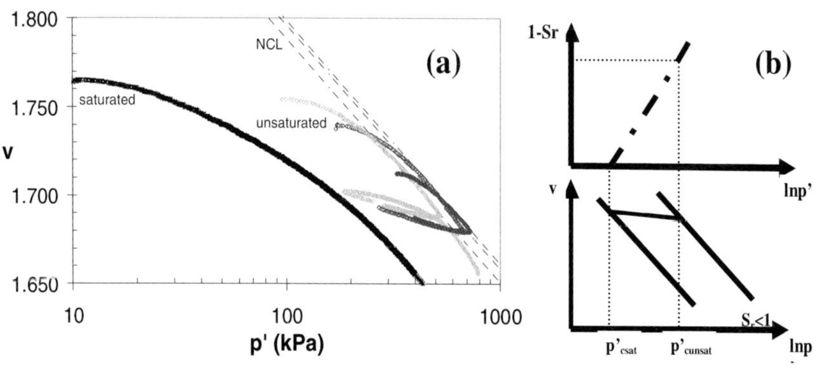

Fig. 2. (**a**) Position of the NCL at various S_r, (**b**) prediction for isotropic state

$$\alpha = A \exp(B n_0) \tag{8}$$

with $A = 6 \times 10^{-09}\,\text{kPa}^{-1}$ and $B = 40$. In Eq. (8), n_0 is the porosity at the end of equalization. Such dependency was chosen to isolate the effect of the imperfect homogeneity of after compaction porosities of the tested specimens, which affect after-equalization values (Vassallo et al. 2006a).

Thus, a single relationship was assumed for S_r variations, without taking into account possible differences between wetting and drying paths. Certainly, this will have to be considered in the continuation of the study.

3 Interpretation of Results at Moderate Strains

In Figure 3 the experimental results of three compression stages (loading and unloading) are compared with model predictions. It is worth noting that, also for most of other compressions, the model predicts an overconsolidated state at the end of equalization, according to experimental data. A good qualitative and quantitative correspondence is found in general.

Figure 4 is relative to two tests including suction cycles.

Test `mp05RC` included a compression at $s = 200\,\text{kPa}$, up to $p - u_a = 200\,\text{kPa}$ and suction cycles between 100 and 400 kPa. The model (Fig. 4) overestimates the plastic strains due to compression and underestimates the strains along the first drying at $s = 400\,\text{kPa}$. This is due to the chosen slope of WRC which, in this case, leads to overestimate the reduction in S_r due to the drying process. Because of this, the model predicts an excessive reversible component of hardening and underestimates plastic strains. The subsequent wetting and drying stages only induce reversible strains, according to experimental data. Better predictions could be probably achieved by taking into account the hydraulic hysteresis of the WRC.

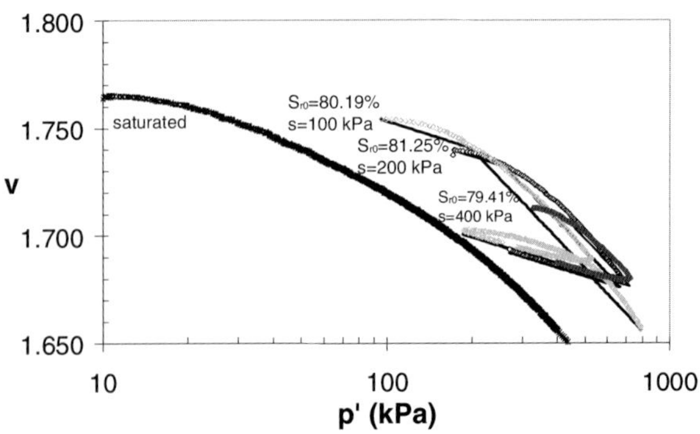

Fig. 3. Comparison between experimental results and model predictions for compression tests carried out at three different suction levels

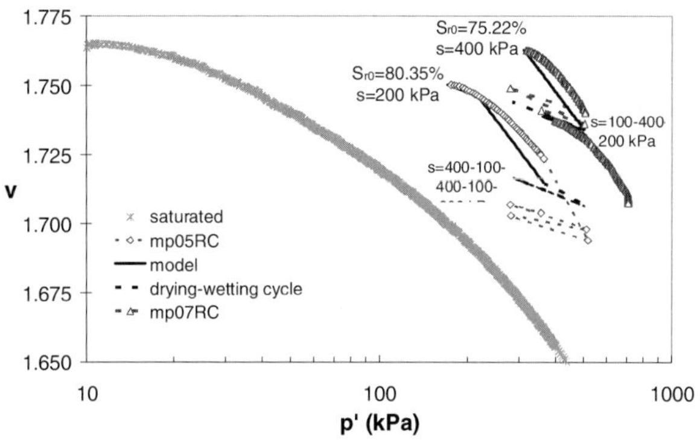

Fig. 4. Comparison between experimental results and model predictions for tests including compression and drying-wetting stages

Test mp07RC included a compression at $s = 400\,\text{kPa}$, up to $p - u_a = 200\,\text{kPa}$ and suction cycles between 100 and 400 kPa. The model predicts normal consolidation at the beginning of the test, while experimental results in the plane $v - p'$ show a light overconsolidation. Predictions relative to the following wetting, drying and compression stages are in good agreement with experimental data.

4 Interpretation of Results at Small Strains

The proposed model can also be used as a framework to interpret the variations of initial shear stiffness G_0 with mean net and suction observed by Vassallo et al. (2006a).

Before proposing such interpretation, it is worth recalling some features of saturated soil behaviour. Initial stiffness depends on the average effective stress p'. Rampello et al. (1994), examining data relative to reconstituted and undisturbed materials, showed that an undisturbed material, thanks to its "structure," has higher values of G_0 than the same material in a reconstituted state (Fig. 5). The effect of this structure is lost as an effect of increasing p'. Furthermore, reconstituted normally consolidated and overconsolidated materials show $G_0 : p'$ curves qualitatively similar to the curves $v : p'$ in the compressibility plane. A linear relationship can be assumed between $\log G_0$ and $\log p'$ for both normally and over-consolidated states, with a lower slope for overconsolidation.

Using same representations as Rampello et al. (1994), G_0 values measured under controlled suction conditions by resonant column tests have been plotted in Fig. 5**a** as a function of the stress p' defined by Eq. (1). Moving from complete saturation to partial saturation induces a translation upwards of experimental curves $G_0 : p'$, similar to that highlighted by Rampello et al. (1994)

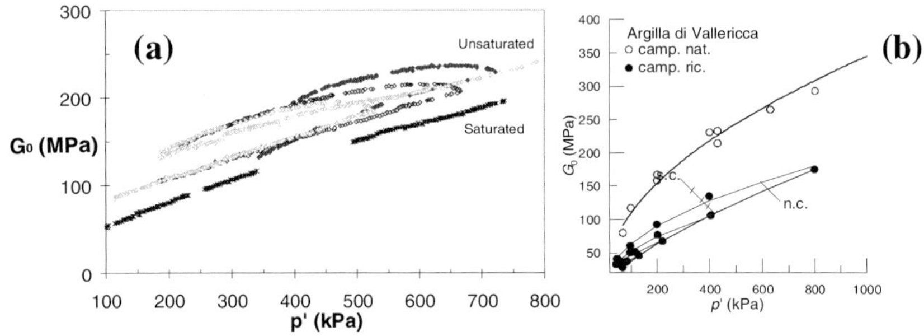

Fig. 5. Comparison: (**a**) saturated-unsaturated compacted soil, (**b**) natural-reconstituted saturated soil

comparing the behaviour of natural and reconstituted clays. Partial saturation seems to provide the solid skeleton with a structure (or cementation) similar to ageing and other natural time effects. In addition, experimental points relative to measurements of G_0 at beginning of the tests show trends similar to that of an overconsolidated saturated soil.

This is in agreement with the observations made above for the end of equalizations: the soil results overconsolidated also in the compression plane $v : p'$. All the above suggests that a reversible phenomenon of "bonding" is associated to a drying path.

A first attempt to interpret this phenomenon was made by considering the hardening predicted by the SSM.

The relationships proposed by Rampello et al. (1994) for saturated soils was extended by including the effects of S_r on void ratio and on p':

$$G_0 = S^* p_r^{(1-n^*+c)} f(e) p'^{(n^*-c)} \tag{9}$$

where

$$f(e) = C/\exp(De) \tag{10}$$

and: $c = m/\Lambda$, $\Lambda = (\lambda - \kappa)/\lambda$, $C = \exp[c(N(S_r) - 1)/\lambda]$, $D = c/\lambda$; p_r is a reference pressure.

Parameters S^*, n^* and m (linked to the position and the slope of $\log G_0 : \log p'$ relationships) were obtained by fitting the saturated soil data (Fig. 6). Resulting values are $n^* = 0.61$, $m = 0.2$, $S^* = 3000$.

It was found that Eqs. (9) and (10) significantly underestimate experimental stiffness of the unsaturated soil. Thus, the extension of saturated relationships to the unsaturated case requires further assumptions. A better agreement can be obtained by supposing that also n^* and S^* depend on S_r. The obtained results are shown in Fig. 6 for the same tests of Fig. 3. It was found that n^* decreases and S^* increases as S_r decreases. The values used in the figure are $n^* = 0.60$ and $S^* \cong 4300$ for $S_{r,0} \cong 85\%$ (Fig. 6**a**), $n^* = 0.59$ and $S^* \cong 4500$ for $S_{r,0} \cong 82\%$, $n^* = 0.59$ and $S^* \cong 4800$ for $S_{r,0} \cong 79\%$ (Fig. 6**b**).

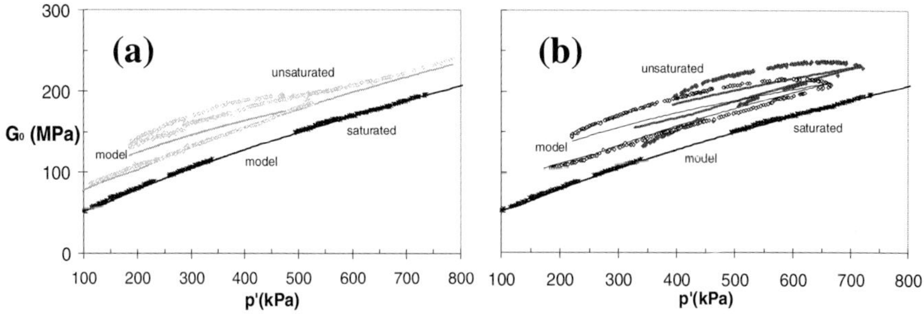

Fig. 6. Comparison between experimental results and model predictions of initial shear stiffness G_0. (**a**) compression test carried out at $s = 100\,\text{kPa}$, (**b**) compression test carried out at $s = 200$ and $400\,\text{kPa}$

5 Conclusions

In this paper it was verified if the comprehensive experimental study on the Po silt by Vassallo et al. (2006a) can be interpreted within the framework of a Single Stress Model (SSM).

Such model can interpret the progressive shift in the plane $v : p'$ of the normal consolidation line as the degree of saturation decreases and, more in general, the influence that S_r has on compressibility.

The effects of partial saturation on the initial shear stiffness G_0 are similar to those ascribable to the structure of a natural soil, compared to the same soil reconstituted. They cannot be justified by simply taking into account the effect that the SSM predicts on the void ratio without significantly underestimating the unsaturated soil stiffness. On the contrary, it is necessary to consider explicitly the effect that S_r has on some parameters describing the relationship $G_0 : p'$.

This is a preliminary study, and some simplifications, such as neglecting the hydraulic hysteresis, will have to be removed in the continuation. However, the results thus far obtained seem encouraging.

References

Houlsby GT (1997) The work input to an unsaturated granular material, Geotechnique 47(1):193–196

Jommi C, di Prisco C (1994) A simple theoretical approach for modelling the mechanical behaviour of unsaturated soils. In: Italian Conference, Il ruolo dei fluidi nei problemi di ingegneria geotecnica, Mondovi, pp 167–188

Rampello S, Silvestri F, Viggiani G (1994) The dependence of small strain stiffness on stress state and history for fined grained soils: the example of Vallericca clay. In: Proc 1st Int Symp on Pre-Failure Defomation of Geomaterials, Sapporo, pp 273–278

Romero E, Vaunat J (2000) Retention curves of deformable clay. Experimental evidence and theoretical approaches in unsaturated soils, Trento, pp 91–106

Tamagnini R (2004) An extended Cam-clay model for unsaturated soils with hydraulic hysteresis, Géotechnique 54:223–228

Van Genuchten MTh (1980) A closed form equation for predicting the hydraulic conductivity of unsaturated soils, Soil Sci Soc Amer J 44(5):892–898

Vassallo R, Mancuso C, Vinale F (2006a) Effects of net stress and suction history on the small strain stiffness of a compacted clayey silt. Can Geotech J (accepted)

Vassallo R, Mancuso C, Vinale F (2006b) Modelling the influence of stress-strain history on the initial shear stiffness of an unsaturated compacted silt, Can Geotech J (accepted)

Independent Roles of the Stress State Variables on Volume–Mass Constitutive Relations

Delwyn G. Fredlund[1] and Hung Q. Pham[2]

[1] Golder Associates Ltd., 145 – 1st Avenue North, Saskatoon, SK., Canada, S7K 1W6 del_fredlund@golder.com
[2] National University of Civil Engineering, 55th Giai Phong Street, Hanoi, Vietnam phamquanghung@gmail.com

Summary. This paper investigates the individual character of the two independent stress tensors and the effect of each on the volume-mass constitutive relations for unsaturated soils. Cases involving soils with varying plasticity and stress history are presented. A model for void ratio (or specific volume) behaviour has been developed within the context of independent stress state variables. Theoretical and experimental results show that changes in the net total stress tensor produce distinctly independent volume change behaviour from changes in matric suction even when the net total stress changes are isotropic. Stress history is reflected differently on the volume change constitutive relations depending upon which stress tensor is changed. Likewise, the water content constitutive relationship is affected in a distinctively different manner depending upon which stress tensor is changed.

Key words: unsaturated, soil, volume, mass, relationship, stress, variable

1 Introduction

Independent stress state variables were introduced in the form of two tensors in the 1970s to describe the stress state for unsaturated soils (Fredlund and Morgenstern 1977). Since that time there have been numerous attempts to develop elasto-plastic models to describe the behaviour of unsaturated soils (Alonso et al. 1990, Wheeler and Sivakumar 1995, Blatz and Graham 2003). Generally, simplistic assumptions have been made with regard to the relationship between water content changes and volume changes in an unsaturated soil.

A volume-mass constitutive model in terms of two independent stress state variables is presented in this paper. The model is capable of predicting volume change and water content independently. The hysteretic nature of the soil-water characteristic curve is also taken into account. The model is proposed for only isotropic condition (i.e., ignores the existence of the shear stress).

2 Terminology

Deformation of a soil mass is directly related to a change in the volume of pores in the soil. The pore-size distribution curve (PDC) of a soil at any stress state provides information regarding both the total volume and the volume of water in the soil. Therefore, the prediction of the PDC is important in a volume-mass constitutive model.

The PDC is a function of soil suction and net mean stress (Pham 2005). In the proposed model, a reference pore-size distribution is first selected. The stress-strain relationship for the soil structure surrounding a pore group is then presented and followed by the calculation for pore groups along the reference pore-size distribution curve.

3 Theory

3.1 Stress State Variables

The proposed model makes use of two stress state variables (Fredlund and Morgenstern 1977): net mean stress $p = (\sigma_1 + \sigma_2 + \sigma_3)/3 - u_a$ and soil suction $\psi = (u_a - u_w)$. The reference stress state is chosen to be a net mean stress = 1 kPa and soil suction = 0 kPa (i.e., equivalent with effective stress = 1 kPa at saturation). The void ratio, e, of the soil is the primarily variable used to represent the overall volume of the soil. The gravimetric water content, w, of the soil is the primary variable used to represent the amount of water in the soil.

3.2 3.2 Reference Pore-Size Distributions

The PDC corresponding to the completely dry condition obtained from an initial slurry soil (i.e., at 10^6 kPa on the initial drying process of the slurry soil) provides a meaningful reference state. In this model, the *reference pore-size distribution curve* of a soil is defined as the PDC corresponding to the completely dry condition of a soil.

A soil has two reference pore-size distribution curves at completely dry conditions (i.e., with respect to drying and wetting suctions). The *reference drying DPC* of a soil provides information regarding the air entry value and the distribution in volume of pores in the soil. Similarly, the *reference wetting PDC* of a soil provides information regarding the water entry value and the distribution in volume of pores in the soil.

3.3 Basic Assumptions

Six assumptions based on the results of previous studies were adopted for the proposed constitutive model.

- *Assumption #1*: A particular pore under consideration in the soil has only two states; namely, i) the pore is filled with water; or ii) the pore is dry.
- *Assumption #2*: Each water-filled pore in the soil also has two indices, namely: i) virgin compression index, and ii) unloading-reloading compression index.
- *Assumption #3*: There are two types of pores: i) compressible pores and ii) non-compressible pores. The compressible pores are relatively large pores and the non-compressible pores are relatively small-interconnected pores.
- *Assumption #4*: Virgin and unloading-reloading compression indices for a pore in the soil are proportional to the volume of the pore at the reference stress state.
- *Assumption #5*: Pores are deformed, and water is absorbed and drained independently.
- *Assumption #6*: Air-filled pores are incompressible.

3.4 Mathematical Formulation

The initial drying SWCC from the slurry of a significant volume change soil can be best-fitted using the following empirical equation:

$$w(\psi) = \left(\left[w_{\text{sat}} - \frac{C_c}{G_s}\log\psi - w_r\right]\frac{a}{\psi^b + a} + w_r\right)\left(1 - \frac{\ln\left[1 + \frac{\psi}{\psi_r}\right]}{\ln\left[1 + \frac{10^6}{\psi_r}\right]}\right) \tag{1}$$

where C_c = virgin compression index of the soil; G_s = Particles specific gravity; w_r = curve-fitting parameter represents the residual water content; w_{sat} = curve-fitting parameter represents the water content at reference stress state; and a, b = curve-fitting parameters. The virgin compression index of the group of pores having air entry value of ψ on the reference pore-size distribution curve can be calculated as follows:

$$C_c^p(\psi) = -\frac{C_c ab\ln(10)\psi^b}{[\psi^b + a]^2}\,. \tag{2}$$

Similarly, the unloading-reloading compression index of the group of pores having air entry value of ψ on the reference pore-size distribution curve can be calculated as follows:

$$C_s^p(\psi) = -\frac{C_s ab\ln(10)\psi^b}{[\psi^b + a]^2}\,. \tag{3}$$

Comparing the strain of a pore that is dried under zero net mean stress and a pore that is dried under a yield stress, p_y, and then dried under a constant net mean stress, p, the equation for the effect of the net mean stress on the air entry value can be written:

$$\frac{\psi_{\text{ae}}}{\psi_{\text{ae}}(p, p_y)} = 1 - \eta\frac{[(C_c - C_s)\log p_y + C_s\log(\psi_{\text{ae}} + p) - C_c\log(\psi_{\text{ae}})]}{3[e_{\text{sat}} - C_c\log\psi_{\text{ae}} - w_r G_s]} \tag{4}$$

where ψ_{ae} = water entry value of the pore having zero yield stress and wetting under zero net mean stress; $\psi_{ae}(p, p_y)$ = air entry value of the pore when yield stress is equal to p_y and drying under a net mean stress of p; η = pore-shape parameter represents the effect of the net mean stress to the change in the diameter of a pore. Similarly, the equation for the effect of the net mean stress to the water-entry value of the pore can be obtained.

The yield stress of the soil structure surrounding a water-filled pore is considered to be maximum effective stress that ever acted on the pore. If a pore is dried under zero net mean stress, the yield stress of the pore is equal to the air entry value of the pore. If a pore is dried under a constant net mean stress of, p, the yield stress of the pore can be calculated:

$$p_y = p + \psi(p, p + \psi_{ae}) \tag{5}$$

where $\psi(p, p+\psi_{ae})$ can be calculated using Eq. (4). When a pore is filled with air, the pore is incompressible (i.e., assumption #6); therefore, yield stress of the soil structure surrounding the pore does not change with net mean stress and soil suction. The two constitutive equations for the volume-mass constitutive surfaces follow the stress paths shown in Fig. 1. The constitutive surfaces correspond to a slurry soil that is initially loaded to a net mean stress, p_0 at zero soil suction and then dried.

Equations have been written for the water content and the void ratio constitutive surfaces (see Pham 2005). The equation for the degree of saturation surface can be derived from the equations for the gravimetric water content surface and void ratio surface.

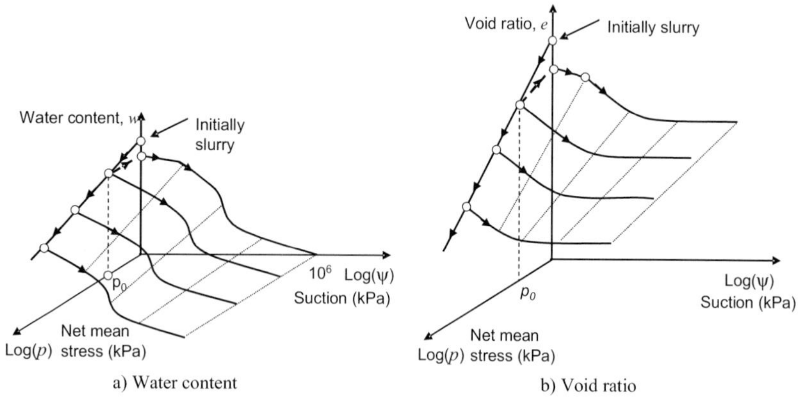

Fig. 1. Schematic illustration of the volume-mass constitutive surfaces of an initially slurred specimen that is dried under various constant net mean stresses

3.5 Hysteresis Model for SWCC

A hysteresis model is developed using two one-dimensional pore-size distribution functions (i.e., wetting and drying pore-size distributions). Scanning curves are horizontal on degree of saturation SWCC plots (similar to the Wheeler et al. (2003) model). This means that each group of pores has a unique relationship between the drying and wetting suction. Therefore, there is a relationship between the wetting reference DPC and the drying reference DPC.

3.6 Model Parameters

The required data for calibration can be described as follows:

- the initial drying SWCC of the initially slurry soil specimen,
- pore-shape parameter, η,
- the parameters for the hysteretic nature of the SWCC of the soil (Pham et al. 2005),
- the compression indices of the soil.

4 Presentation of the Model Prediction

The application of the proposed volume-mass constitutive model is presented for an artificial silt. The soil properties of the artificial silt are shown in Table 1. The initial drying SWCC for the silt is shown in Fig. 2.

Table 1. Soil properties for the artificial silt

Soil-water characteristic curve				Hysteresis parameters			Compression indices		
w_{sat}	a	b	w_r	D_{SL}	R_{SL}	β	C_c	C_s	G_s
0.45	200,000	2.5	0.08	0.35	1.5	0.1	0.2	0.04	2.7

ht The predicted shrinkage curve for the artificial silt shown in Fig. 3 has a reasonable shape. At high water contents (i.e., almost 100% degree of saturation), the shrinkage curve is a 45 degree line. The shrinkage curves are approximately horizontal at low water contents.

Model predictions for the artificial silt when following a complex stress path (Fig. 4) are presented in Fig. 5. Figure 5**a** shows that the model can take into account plastic volume change along the initial drying process (i.e., drying the soil to 100 kPa). The collapsible behavior of the unsaturated silt during a

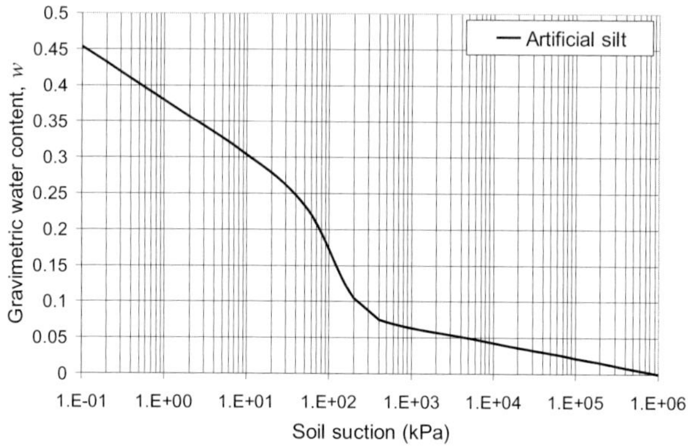

Fig. 2. Initial drying Soil-Water Characteristic Curve for artificial silt

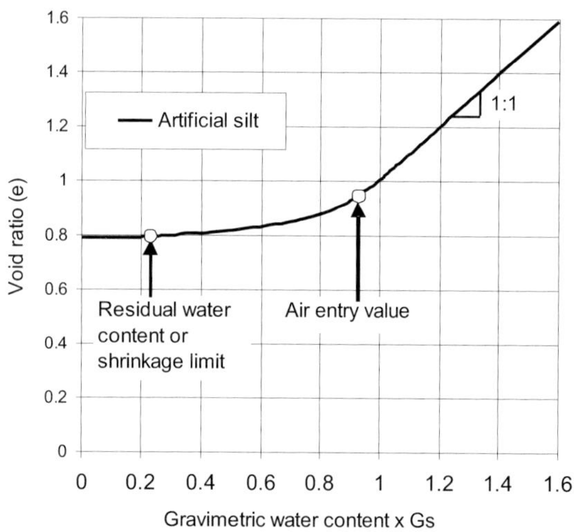

Fig. 3. Predicted shrinkage curve for artificial silt

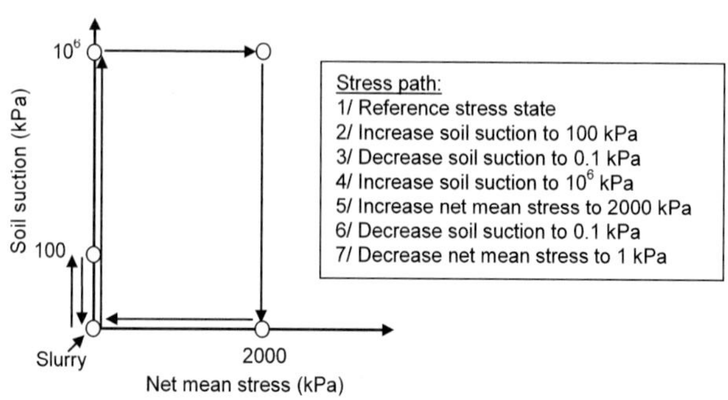

Fig. 4. Stress path followed for artificial silt

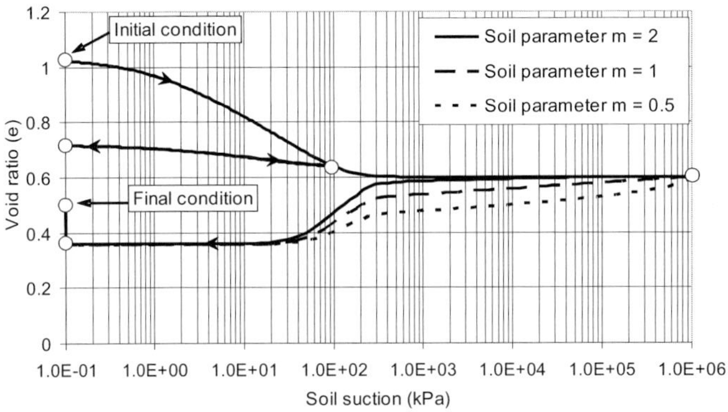

a) Void ratio

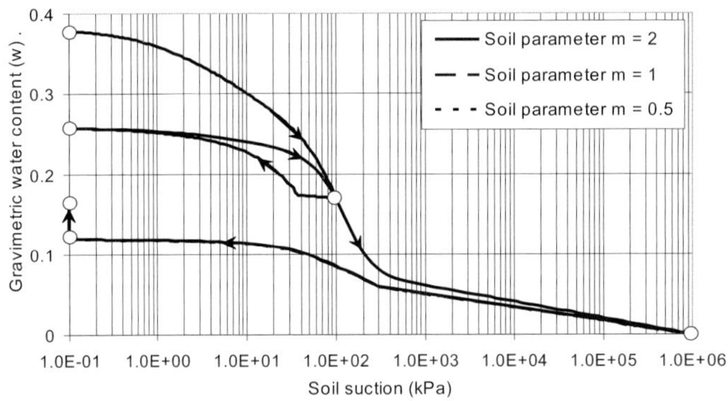

b) Gravimetric water content

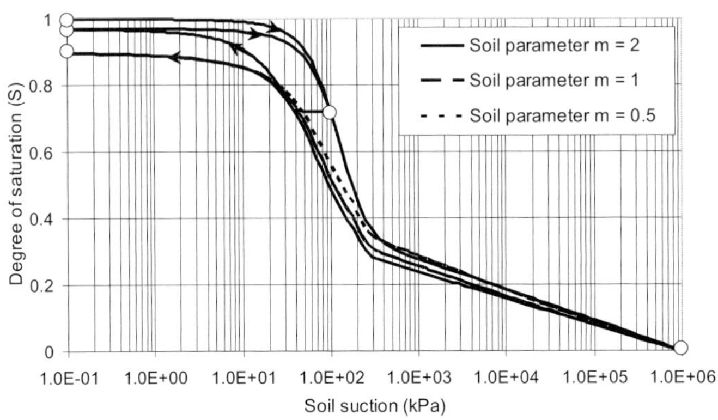

c) Degree of saturation

Fig. 5. Plots of predicted void ratio, gravimetric water content and degree of saturation for artificial silt wetting process can also be seen in Fig. 5**a**. This mechanical behavior can be observed from many experimental results (Matyas and Radhakrishna 1968, Alonso et al. 1990). The soil parameter m controls the collapsible behavior of the soil. The smaller the value for the soil parameter, m, the earlier the collapse occurs in the soil during the wetting process.

The hysteresis in the gravimetric water content shown in Fig. 5**b** results from: i) plastic volume change and ii) the hysteretic nature of the SWCC. In general, mechanical behaviors of the soil in the model agree well with experimental results of actual soils (Pham 2005).

5 Conclusions and Recommendations

The proposed volume-mass constitutive model is capable of: 1) predicting both volume and water content at all stress states corresponding to a wide variety of stress paths; 2) taking into account the hysteretic nature of the soil-water characteristic curve; and 3) predicting both swelling and collapsible behavior of an unsaturated soil. The model can predict volume-mass constitutive relationships that are stress path dependent. The prediction results appear to be consistent with observed laboratory data.

References

Alonso EE, Gens A, Josa A (1990) A constitutive model for partially saturated soils, Geotechnique 40(3):405–430

Blatz JA, Graham J (2003) Elastic-plastic modelling of unsaturated soil using results from a new triaxial test with controlled suction, Géotechnique 53(1):113–122

Fredlund DG, Morgenstern NR (1977) Stress state variables for unsaturated soils, J Geotech Eng Div ASCE (GT5) 103:447–466

Matyas EL, Radhakrishna HS (1968) Volume change characteristics of partially saturated soils, Geotechnique 18(4):432–448

Pham QH (2005) A volume-mass constitutive model for unsaturated soils. PhD Thesis dissertation, University of Saskatchewan, Canada, 404 pp

Pham QH, Fredlund DG, Barbour SL (2005) A study on the hysteresis models for soil-water characteristic curves, Can Geotech J 42(6):1548–1568

Wheeler SJ, Sivakumar V (1995) An elasto-plastic critical state framework for unsaturated soil, Géotechnique 45(1):35–53

Wheeler SJ, Sharma RJ, Buisson MSR (2003) Coupling of hydraulic hysteresis and stress-strainbehaviour in unsaturated soils, Géotechnique, 53(1):41–54

The Soil Water Characteristics of Two-Component Sand Mixtures

Emőke Imre[1], Imre Laufer[1], Quoc Phong Trang[1], János Lőrincz[2], Kálmán Rajkai[3], Tibor Firgi[4], and Gábor Telekes[4]

[1] Geotechnical Research Group of the Hungarian Academy of Sciences at BME, Budapest, Hungary
imreemok@hotmail.com, bincy@freemail.hu, tqp0322@gmail.com
[2] Gradex, Budapest, Hungary lorincz@gradex.hu
[3] Research Institute for Soil Science and Agricultural Chemistry, Hungarian Academy of Sciences, Budapest, Hungary krajkai@rissac.hu,
[4] St. István University, Ybl Miklós School of Engineering, Budapest, Hungary
Firgi.Tibor@ymmfk.szie.hu, Telekes.Gabor@ymmfk.szie.hu

Summary. During the ongoing research the water retention curves of three sand fractions and 12 continuous and gap-graded 2-component sand mixtures is measured. Due to the unexpectedly long equalization times complete data sets are not available yet. The existing data are used to test the newly set up sand box, an a priori grading curve–water retention curve model, the van Genuchten model and the Fredlund–Xing model. Comparing the old and the new boxes, first results indicate a constant bias. The a priori model is partly supported by the measured data indicating that the present grading curve determination method is not sophisticated enough. According to the results of the model fitting, the van Genuchten model is slightly better for the tested sands than the Fredlund–Xing model and, the fit is better for short data sets than for the whole SWCC. According to computed van Genuchten permeability functions indicates that the permeability decreases more rapidly for the finer fractions than for the coarsest fraction explaining the long equalization times.

Key words: unsaturated soil functions, sand, measurement, inverse problem

Introduction

The most important unsaturated soil functions are (i) the water-retention curve or soil water characteristic curve (SWCC), (ii) the permeability function and, (iii) the shear strength function. The measurement of the unsaturated soil functions is time consuming. Several methods are available for estimating the permeability function and the shear strength function from the water-retention curve. However, the knowledge is limited concerning the relationship between the grading curve and the soil water characteristic curve (SWCC). The aim of an ongoing research (Imre et al. 2003, 2005, 2006) is to produce soil property functions from the grading curve using the concept of the grading entropy.

In this paper some preliminary results concerning three sand fractions and 12 two-component continuous or gap-graded sand mixtures are presented. Only the initial part of the SWCC determination results are published in this paper due to the unexpectedly long equalization times. They were used to test a newly set up SWCC measuring box, the suggested a priori mixture retention model, the van Genuchten and the Fredlund–Xing models.

1 Grading Entropy Concept

1.1 Grading Entropy Formulae

The grading curve is a statistical distribution curve where the probability is given in the function of the diameter (d) of the particles. The grading entropy is an application of the statistical entropy based on the grain size distribution curve (Lőrincz 1990, 1993a,b, Lőrincz et al. 2005, Imre 1995). For the application of the statistical entropy concept, the statistical cells are selected as the "fractions." The fractions are originally defined as the grains "passing through one mesh, but retained by the next." This definition is extended as follows. The limiting d values for the i-th fraction in terms of $d_{\min}$ are as follows:

$$2^j d_0 \geq d > 2^{j-1} d_0 \,. \tag{1}$$

The minimum of the grain diameter ($d_{\min}$) is specified as 2^{-22} mm. The fractions are numbered by the increasing integers i (Table 1). Considering a grading curve, N is used for the number of the fractions between the finest and coarsest ones and, n is the number of the fractions with non-zero relative frequency. The fractions may be numbered in relative systems, too (e.g. Table 2). Using this notation, it holds for the relative frequencies x_i:

$$\sum_{k=1}^{N} x_k = 1, \qquad x_k \geq 0 \,. \tag{2}$$

The grading entropy of the soil (S) can be split into two parts according to:

Table 1. Eigen-entropy S_{0i} of the fractions

Fraction		24	23	22	
d (mm)	...	4–8	2–4	1–2	...
$S_{0i} = i$		24	23	22	

Table 2. Fractions

Fraction	Simplified notation	Grain size d (mm)
17	7	0.03–0.06
18	6	0.06–0.125
19	5	0.125–0.25
20	4 (A)	0.25–0.50
21	3 (B)	0.50–1.0
22	2 (C)	1.0–2.0
23	1	2.0–4.0

$$S = \Delta S + S_0 \,. \tag{3}$$

The entropy increment (ΔS) is given by:

$$\Delta S = -\frac{1}{\ln 2} \sum_{i=1}^{N} x_i \ln x_i \,. \tag{4}$$

The base entropy (S_o) is given by:

$$S_o = \sum_{i=1}^{N} x_i S_{oi} \tag{5}$$

where S_{0i} are the eigen-entropy values of the fractions shown in Table 1.

1.2 Entropy Diagram

The normalised form of the base entropy, the relative base entropy (A) is defined as:

$$A = \frac{S_o - S_{o\,\min}}{S_{o\,\max} - S_{o\,\min}} \tag{6}$$

where $S_{o\,\max}$ and $S_{o\,\min}$ are the eigen-entropy of the largest and the smallest fractions in the mixture.

The "normalised" entropy increment (B):

$$B = \frac{\Delta S}{\ln N} \,. \tag{7}$$

Any grading curve can be represented by a point either in an $N-1$ dimensional simplex or in two dimension, in terms of the normalised entropy coordinates A and B. The map between these two representations is continuous, the image is compact (Fig. 1).

The relative base entropy (A) may vary from 0 to 1, The normalised entropy increment (B) may vary from 0 to $1/\ln 2$. The normalised entropy increment (B) has a maximum and a minimum if the values of A and

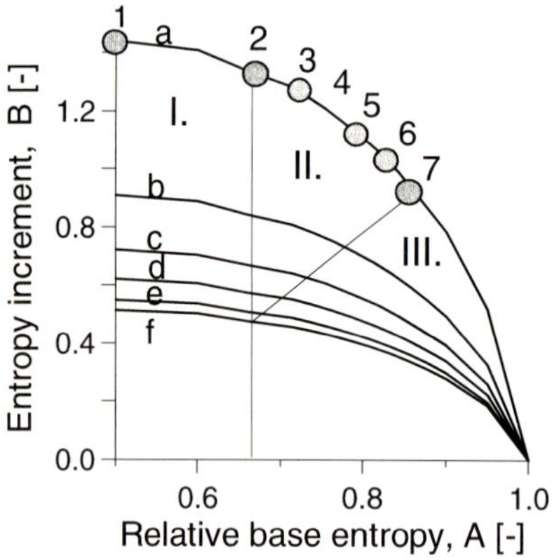

Fig. 1. Half of the simplified entropy diagram indicating the grain structure stability domains (I: piping, II: stable, III: sufosion), lower bounding entropy increment lines (**a** to **f** for $N = 2$ to 7) and, maximum entropy points (2 to 7 for $N = 2$ to 7)

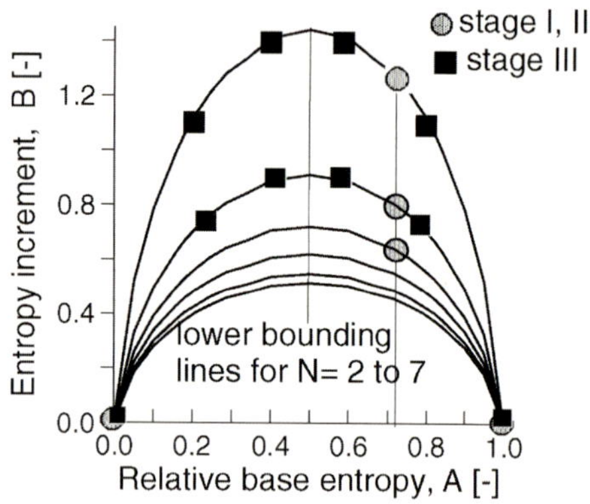

Fig. 2. Soils tested in the simplified entropy diagram (the first, the second and the third part of the research)

N are specified. The maximum line of B is slightly dependent on N for $0 < A < 0.5 < A < 1$ but this is practically negligible.

The minimum bounding line of B in the simplified entropy diagram is defined by such gap-graded grading curves where all x_i are zero except x_1 and x_N. The maximum of these curves at $A = 0.5$ is $B = 1/\ln N$ being dependent of N. These two curves (the maximum and the minimum lines of B) are symmetric to the line of $A = 0.5$ and coincide if $N = 2$.

1.3 Some Test Results in the Entropy Diagram

Two basic types of the soil structures can be distinguished by representing the grading curves as a point in the entropy diagram (Lőrincz et al. 2005). If A is less than 2/3, the coarser particles "float" in the matrix of the finer ones, and the soil structure is unstable. If A is greater than 2/3 the coarser particles form a permanent skeleton, the structure is stable. The dry density in the loosest possible or e_{max} state is a maximum in some sense when the point is situated at the $A = 2/3$ line (Fig. 1). The related grading curves are the best for concrete production and are not sensitive to segregation.

Table 3. Mixtures used in the first part of the research

The number of the non-zero fractions between the finest and coarsest ones, n [–]	The number of the fractions between the finest and coarsest fractions, N [–]	A [–]	Fractions in the mixture
2	2	2/3	1–2, 2–3, 3–4, 4–5, 5–6, 6–7
3	3	2/3	1–2–3, 2–3–4, 3–4–5, 4–5–6, 5–6–7
4	4	2/3	2–3–4–5, 3–4–5–6, 4–5–6–7
5	5	2/3	1–2–3–4–5, 2–3–4–5–6, 3–4–5–6–7
6	6	2/3	1–2–3–4–5–6, 2–3–4–5–6–7
7	7	2/3	1–2–3–4–5–6–7

Table 4. Mixtures used in the second part of the research

n [–]	N [–]	A [–]	Fractions in the mixture
2	2	2/3	1–2, 2–3, 3–4
2	3	2/3	1–3, 2–4
2	4	2/3	1–4

2 Materials and Methods

2.1 Fractions and Mixtures

The fractions are numbered consecutively from the coarsest one (Table 2). In the first stage of the research (Imre et al. 2003) seven fractions and 21 continuous mixtures with grading curve related to the optimal point (maximum B at $A = 2/3$, Table 3, Fig. 1) were tested.

The soils used in the second part of the research (Imre et al. 2005) are four fractions and six two-component continuous and, gap-graded mixtures (with $A = 2/3$, Table 4, Fig. 1).

In this phase of the research the water retention curves of three sand fractions and 12 continuous and gap-graded two-component sand mixtures with various composition (A = 0–1, Table 5 and Fig. 2). The fraction 2, 3, and 4 were renamed as C, B and A, respectively (Table 1).

Table 5. Mixtures used in the third part of the research

n [-]	N [-]	A [-]	Fractions in the mixture
2	2 or 3	0.2; 0.4; 0.6; 0.8	2–3, 3–4, 2–4 or C–B, C–A, B–A

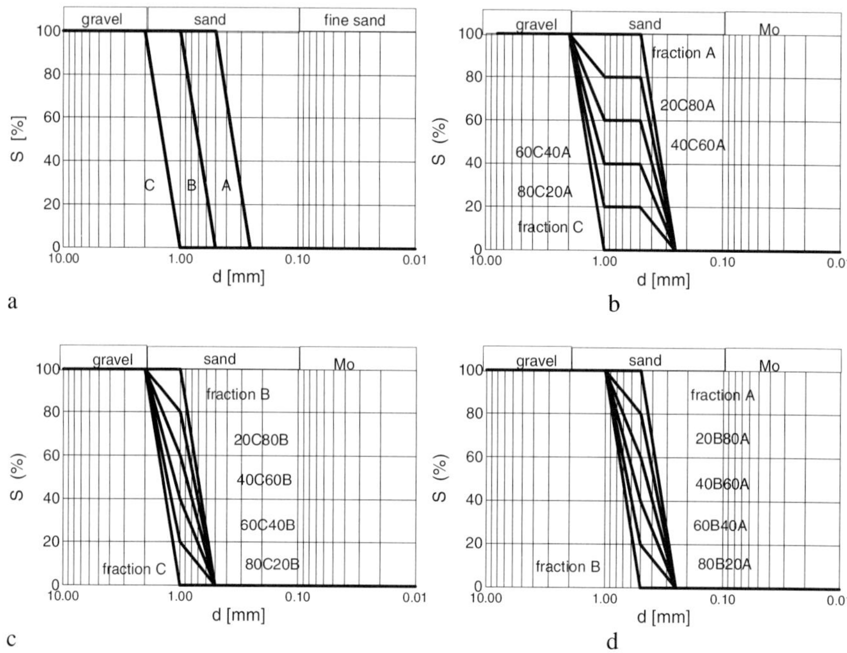

Fig. 3. The grading curve of the soils used in this part of the research. (**a**) The fractions, (**b**) B–A mixtures, (**c**) C–A mixtures, (**d**) C–B mixtures

2.2 Measurements

The method of the Soil Science Institute (Várallyay 1973, Rajkai 1993) was applied in the first stage of the study. In the smaller suction range of $u_a - u_w \leq 50\,\mathrm{kPa}$ sand boxes were used and, the suction was applied by water pressure decrease. The load steps were fixed for each sand box to avoid any disturbance of the sand layer (Table 6). In the greater suction range ($u_a - u_w > 50\,\mathrm{kPa}$) pressure membrane extractors were used.

Table 6. The load steps in low suction range (*bold*: original load step, *, **: new load steps in 2005 and 2006)

Load (cm)	Semi-permeable membrane	Suction load
1, **2.5**	sand	gravitational
4*, 6**, 7*, 8**, **10**, 10**, 13**, 15*, 16**, 20*, 20**, 23**, 26**, 29**, **31.5**, 32**, 35**	fine sand	gravitational
100, **200**, **501**	kaolinit	gravitational

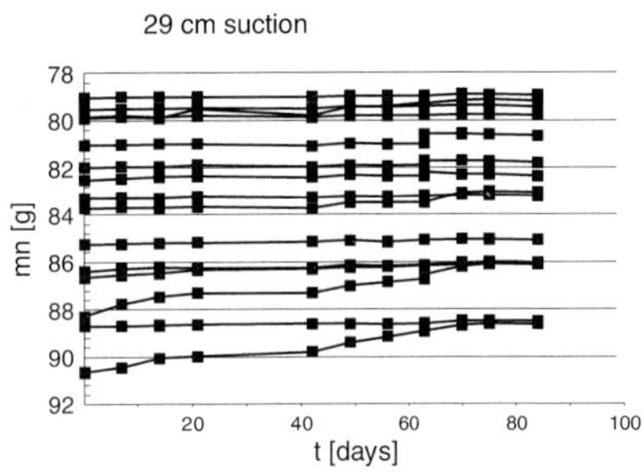

Fig. 4. Equalization during a stage

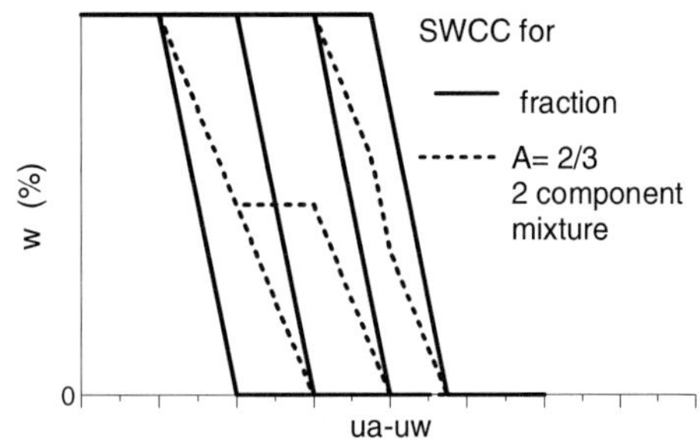

Fig. 5. Model for $A = 2/3$ continuous mixtures using trilinear non-overlapping and overlapping fraction water retention curves

In the second and third stage of the study the suction steps used in the Institute for Soil Science were extended applying a new sand box together with the old ones. Four/twelve new steps were included in the low suction range with a separate sand box (Table 6) in 2005/2006 where the load was varied. In the third stage of the study only the new box was used.

The samples were prepared in the loosest possible dry state then were saturated. Each test sample was doubled. The dry density ρ_d of the samples is measured at the end of the test. No antimicrobial agent was applied.

Originally the stage duration was generally about 7–21 days. In the third part of the research it was realized that much longer load steps are needed. According to the time versus measured mass curve results shown in Fig. 4, it was found that more than two month were needed when for larger suctions instead of 2–3 weeks. That is why half of the double soil samples were open the paper is written on the basis of these data. The measurement of the remainder samples is still going on.

2.3 Model Fitting, Permeability Function

Two following two models were fitted on the data measured for the B–C mixtures with the method described in Imre (1996): the van Genuchten (1980) water retention curve equation:

$$w = w_r + \frac{w_s - w_r}{(1 + [a(u_a - u_w)]^n)^m}$$

and, the Fredlund–Xing (1994) water retention curve equation:

$$w = w_r + \frac{w_s - w_r}{\{\ln[e + [a(u_a - u_w)]^n]\}^m}$$

where w_r, w_s, a, n, m are the model parameters, $u_a - u_w$ is suction. Using the identified van Genuchten model parameters, the corresponding permeability function was computed:

$$k_r = \frac{\{1-(a(u_a-u_w))^{n-1}[1+(a(u_a-u_w))^n]^{-m}\}^2}{[1+(a(u_a-u_w))^n]^{m/2}}\,. \tag{8}$$

The following merit function was minimised:

$$F(\mathbf{p}) = \frac{\sqrt{\sum_{i=1}^{L} w_{me}(u_i) - w(u_i,\mathbf{p})}}{L \max_i (w_{me}(u_i))} \tag{9}$$

where subscript me is measured, $\mathbf{p}$ is parameter vector consisting of the model parameters $[w_r, w_s - w_r, 1/a^n, n, m]$ and u_i, $(i = 1, \ldots, L)$, are the applied suction load steps.

The vector $\mathbf{p}_{\min}$ related to the global minimum of the merit function is reliable if it is a unique solution and, if its error is within the range of the parameters (geometrically the merit function has a "nice" or distinctly determinable and "deep" single global minimum point). The reliability of the solution of the inverse problem was tested as follows. The standard deviation of parameters was analytically determined using the linearized form of the model (Press et al. 1986). The parameter error was geometrically interpreted by representing the merit functions along the so-called minimal sections, too (Imre 1996).

2.4 A Priori Model for the Mixture SWCC

It is assumed that the water retention curve of the mixture (w_m) is the linear (convex) combination of the water retention curves of the fractions (w_k):

$$w_m(u_a - u_w) = \sum_{k=1}^{N} c_k w_k(u_a - u_w)\,. \tag{10}$$

This model is illustrated in Fig. 5 using the simplest possible fraction water retention curves for the $A = 2/3$ mixtures, assuming that all the c_k $(k = 1, \ldots, N)$ are equal to each-other in this case. The steep part of the predicted mixture water retention curves is linear or, it contain parallel linear lines. The shape of the water retention curve of the mixtures is different for overlapping and not overlapping fractions. It follows from this model that the air entry value and the residual value of the mixture agree with the ones of a fraction for any value of A.

3 Results and Discussion

3.1 SWCC Measurement

According to the preliminary results (Fig. 6) there is a shift between the previously and the newly measured fraction water retention curves. This shift

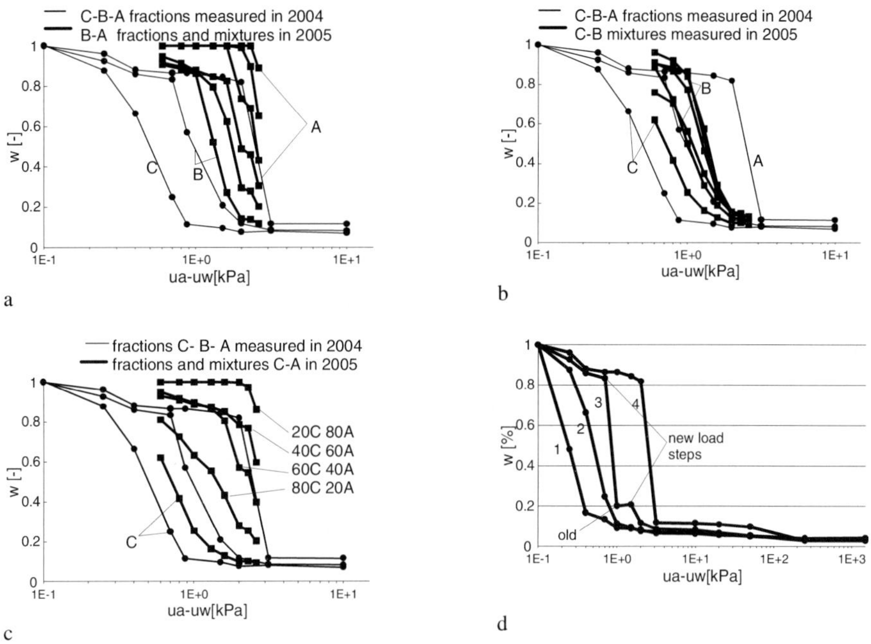

Fig. 6. (**a**) Preliminary B–A mixture water retention curves. (**b**) Preliminary C–A mixture water retention curves. (**c**) Preliminary C–B mixture water retention curves. (**d**) Discrepancy first observed in Imre et al. (2005)

Table 7. Approximate threshold suction data for fractions 1–4

	Air entry suction [kPa]		Residual suction [kPa]	
	2005	2006	2005	2006
1	0.25			
2 (C)	0.4	0.9	0.7	1.4
3 (B)	0.7	0.85	2	21
4 (A)	2		3	

is presented in terms of the approximate air entry suction values and the approximate residual suction values in Table 7.

The measurement was made in different sand boxes and at the different times. The former effect is probably more important since the shift was observed previously, too (Fig. 6**d**). The old boxes and the new box were built of different materials. Further research is needed in this respect.

3.2 Model Fitting

Only the initial part of the measured SWCC concerning the mixtures BC where evaluated by model fitting (Figs. 7–10, Tables 8–12).

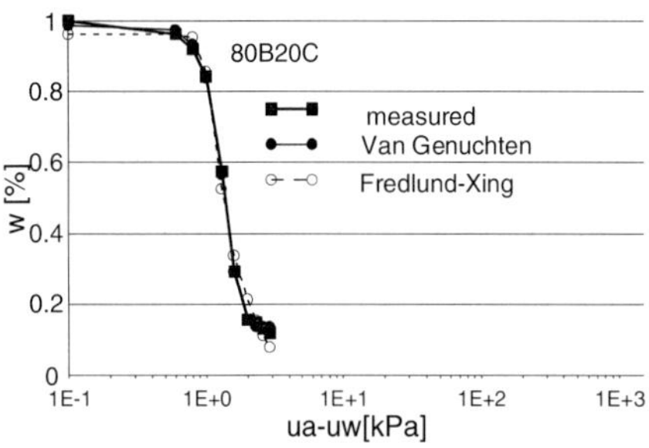

Fig. 7. Typical result for the measured and fitted data, 80%B – 20%C mixture

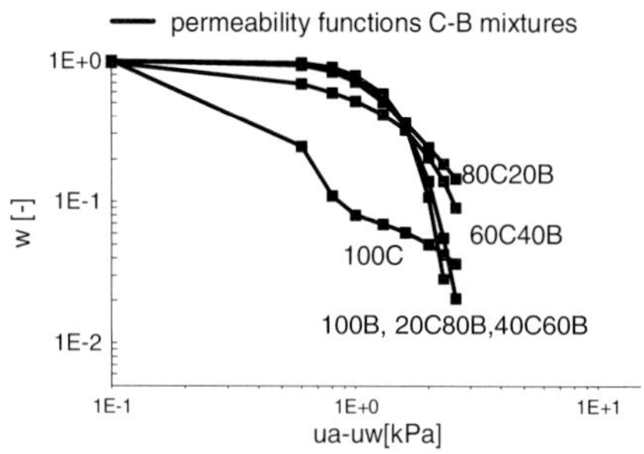

Fig. 8. Permeability functions for the B–C mixtures (determined with the van Genuchten model)

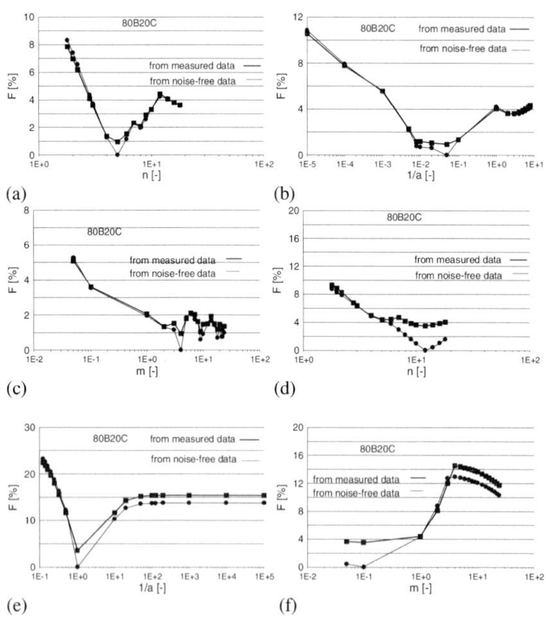

Fig. 9. Minimal sections of the merit function for the non-linearly dependent parameters in the case of the van Genuchten model ((**a**)–(**c**)) and the Fredlund–Xing model ((**d**)–(**f**))

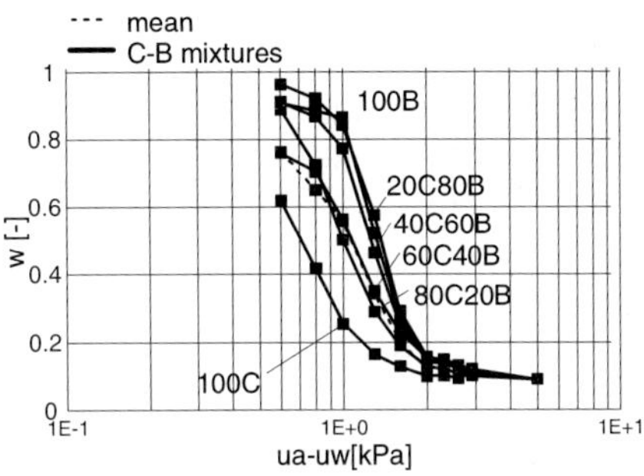

Fig. 10. Mean SWCC for the preliminary C–B mixture water retention curves

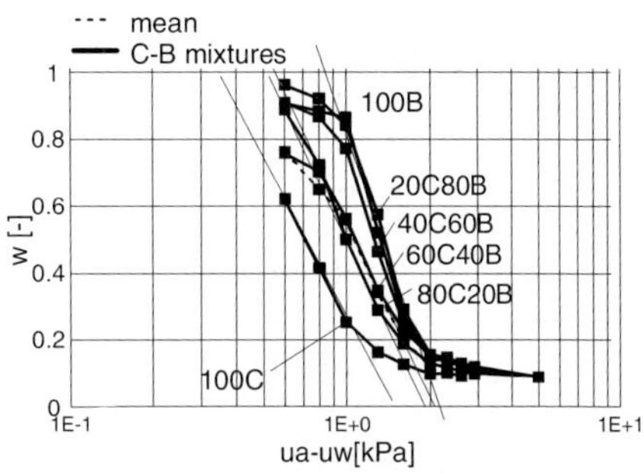

Fig. 11. Approximate air entry and residual suctions concerning the preliminary C–B mixtures

Table 8. Identified parameters and fitting error [%] for the Fredlund–Xing (1994) equation

Soil	$w_r/w(0)$	$(w_s - w_r)/w(0)$	a^n [kPa]$^{-n}$	n	m	Fitting error F [%]
100B	−2.74	3.69	1.00	14.00	0.10	3.76
20C80B	−2.97	3.94	1.00	12.00	0.10	3.53
40C	−0.07	1.05	1.00	6.00	1.00	2.95
60C40B	−0.02	0.99	1.00	2.80	2.00	3.04
20B80C	0.10	0.92	1.00	4.00	3.00	0.69
100C	0.11	0.93	1.00	2.80	7.00	1.21
Mean	−0.93	1.92	1.00	6.93	2.20	2.53

Table 9. Identified parameters and fitting error [%] for the van Genuchten (1980) equation

Soil	$w_r/w(0)$	$(w_s - w_r)/w(0)$	a^n [kPa]$^{-n}$	n	m	Fitting error F [%]
100B	0.12	0.85	0.05	5.00	4.00	2.78
20C80B	0.13	0.85	0.05	5.00	4.00	0.96
40C	0.13	0.85	0.02	4.00	16.00	1.70
60C40B	0.12	0.89	0.10	2.20	8.00	2.15
20B80C	0.08	0.91	1.00	4.00	1.00	1.54
100C	0.10	0.90	5.00	4.00	1.00	1.23
Mean	0.11	0.87	1.04	4.03	5.67	1.73

Table 10. Coefficient of variation [–] for the parameters identified with the van Genuchten (1980) equation

Soil	$w_r/w(0)$	$(w_s - w_r)/w(0)$	a^n	n	m
100B	0.75	0.10	0.30	0.18	0.27
20C80B	1.23	0.18	0.06	0.20	0.40
40C	0.42	0.06	0.15	0.12	0.14
60C40B	0.58	0.08	0.15	0.21	0.14
20B80C	0.63	0.05	0.15	0.14	0.10
100C	0.39	0.04	0.13	0.09	0.08
Mean	0.67	0.09	0.16	0.15	0.19

Table 11. Coefficient of variation [–] for the parameters identified with the Fredlund–Xing (1994) model

Soil	$w_r/w(0)$	$(w_s - w_r)/w(0)$	a^n	n	m
100B	0.04	0.03	0.41	1.02	0.45
20C80B	1.23	0.18	0.06	0.20	0.40
40C	1.36	0.09	0.04	0.13	0.08
60C40B	4.92	0.10	0.08	0.18	0.12
20B80C	1.59	0.06	0.11	0.04	0.02
100C	0.35	0.04	0.02	0.05	0.04
Mean	1.58	0.08	0.12	0.27	0.19

Table 12. Comparing mean fitting errors and coefficient of variations using the short and the previous long data of Imre et al. (2006)

SWCC data set	Model	Fitting error	Parameter				
			w_r	$w_s - w_r$	a^n	n	m
Short	van Genuchten	1.73	0.67	0.09	0.16	0.15	0.19
Long	van Genuchten	4.00	1.72	0.16	0.40	1.19	0.62
Short	Fredlund–Xing	2.53	−0.53	0.08	0.12	0.27	0.19
Long	Fredlund–Xing	3.57	4.41	0.16	0.08	0.18	0.17

According to the results, the agreement between the measured and computed sand data was slightly better for van Genuchten model than for the Fredlund–Xing model. This result is in accordance with some previous observations indicating that the van Genuchten (1980) model is the best for sands, the Fredlund–Xing (1994) model is the best for clays (Agus et al. 2003, Imre et al. 2006).

The fit was better now using the initial part of the curve only than previously using the whole curve for similar soils (Imre et al. 2006, Table 12). This kind of experience may be encountered in the case of other inverse problems, too where "long" and "short" data series are used (e.g. Imre 2002). This can probably be explained by the different contribution of the model errors.

The error of the identified residual water content parameter was the largest for both models in accordance with the general experience (e.g. Groenevelt and Grant 2004). The identified values were surprisingly realistic for the van Genuchten model. Parameters m and n indicated a relatively large error by both the geometrical and the analytic methods. It can be noted that the error of the identified parameters was less now than previously when the whole curve was fitted for similar sand mixtures (Imre et al. 2005, Table 12).

The shape of the corresponding van Genuchten permeability functions was fairly good and it was interesting that the k value decreased faster for the finer fraction B than the coarser C probably due to the fact that the steep part of the SWCC had different slope.

3.3 Mixture Model

In the previous parts of the research $A = 2/3$ mixtures were tested only. Results were approximate since the load steps were too large. These approximate data supported the SWCC mixture model in terms of the slope of the steep part. However, it was found that, the initial water content $w(0)$ was always larger for the fractions than for the mixtures since the initial water content $w(0)$ was dependent on the initial dry density ρ_d:

$$w(0) = 1 - \frac{\rho_d}{\rho_s}. \tag{11}$$

The initial dry density ρ_d was less for the mixtures than for the fractions (Imre et al. 2005).

In this part of the research $0 < A < 1$ two-component mixtures are tested. The mean fraction SWC curves and, the approximate air entry and the residual suction values of the mixtures were constructed for the B–C mixtures in Figs. 10–11. The mean fraction water retention curve is in good agreement with the ones of those mixtures which are the closest to the optimal $A = 2/3$ mixture (Fig. 10). However, the approximate air entry and the residual values of the mixtures differ from the ones of the fractions (Fig. 11, intersection of the slope linear lines and the $w = 0$ or $w = 1$ lines, respectively) not supporting the model at the first glance.

An interesting outcome of the measured results is that the SWCC of the A–B and A–C mixtures contains constant-valued parts. Applying the a priori model, this result can be explained by "gap-graded" grading curves. However, the grading curve is gap-graded for the AC mixtures only (Figs.3 and 6). This result indicates that the distribution of the diameter d of the grains or of the pores within a fraction is possibly not uniform.

4 Conclusions

During the ongoing research the water retention curves of three sand fractions and 12 continuous and gap-graded two-component sand mixtures with various composition ratios (and, as a result, with various A values) are intended to be measured with a new sand box.

A bias was observed between the previously and the newly measured fraction water retention curves (Fig. 6) which was observed previously once, too. The old sand boxes and the new sand box have been built of different materials. The shift can probably partly be attributed to the threshold gradient effect (De Gennaro et al. 2002), partly to the error of the outflow control.

First results indicate longer equalization time than it was expected. The very long stage duration times (about three months) observed for larger suction loads can probably be attributed to the fact that the permeability decreases by a very large amount for the fractions A and C (Fig. 4). Because of this, complete data set especially for the finest fraction A are not available as yet.

The van Genuchten model and the Fredlund–Xing model were fitted on the initial part of the B–C data. The main findings were as follows in accordance with the results of other research in literature.

(i) The fit was slightly better for the former than the latter model (Agus et al. 2003, Imre et al. 2006).
(ii) The fit was better now using the initial part of the curve only than previously using the whole curve for similar soils (Imre et al. 2006, Table 12). This kind of experience may be encountered in the case of other inverse problems, too (Imre 2002).
(iii) The error of the identified residual water content parameter was the largest for both models in accordance with Groenevelt and Grant (2004). The residual water content parameter identified with the former model was realistic similarly to some previous results (Agus et al. 2003, Imre et al. 2006).

The existing data partly support the a priori water retention curve model (i.e. the mixture retention curve is the convex combination of the ones of the fractions). An interesting outcome of the research is as follows. The measured water retention curve may contain a constant part even though in the case when the grading curve is gap-graded. This indicates that the distribution of the diameter d of the grains or of the pores within a fraction is possibly not uniform. Further research is suggested on the determination of the grain and pore size distribution curves of the fractions.

References

Agus SS, Leong EC, Schanz T (2003) Assessment of statistical models for indirect determination of permeability functions from soil–water characteristic curves,

Geotechnique 53(2)279–282
De Gennaro V, Cui YJ, Delage P, De Laure E (2002) On the use of high air entry value porous stones for suction control and related problems, In: Jucá JFT, de Campos TMP, Marinho FAM (eds) Unsaturated Soils. Proc 3rd Int Conf on Unsaturated Soils (UNSAT 2002), Recife, Brazil. Lisse: Swets & Zeitlinger 2:435–440
Fredlund DG, Xing A (1994) Equations for the soil-water characteristic curve, Can Geotech J 31:521–532
Groenevelt PH, Grant CD (2004) A new model for the soil–water retention curve that solves the problem of the residual water contents, Eur J Soil Sci 55:479–485
Imre E (1995) Characterization of dispersive and piping soils. In: Proc XI ECSMFE, Copenhagen 2:49–55
Imre E (1996) Inverse problem solution with a geometrical method. In: Proc 2nd Int Conf on Inverse Problems in Engineering, Le Croisic, France:331–338
Imre E (2002) Evaluation of "short" dissipation tests. In: Proc 12th Danube-European Conference:499–503
Imre E, Rajkai K, Genovese R, Jommi C, Lőrincz J, Aradi L, Telekes G (2003) Soil water-retention curve for fractions and mixtures. Proc of UNSAT-ASIA, Osaka:451–456
Imre E, Havrán K. Lőrincz J, Rajkai K, Firgi T, Telekes G (2005) A model to predict the soil water characteristics of sand mixtures. In: Int Symp on Advanced Experimental Unsat Soil Mech, Trento, June 27–29
Imre E, Rajkai K, Firgi T, Trang QP, Telekes G (2006) Closed-Form Functions for the Soil Water-Retention Curve of Sand Fractions and Sand Mixtures. In: 4th Int Conf On Unsaturated Soils, Arizona:2408–2419
Lőrincz J (1990) Relationship between grading entropy and dry bulk density of granular soils, Periodica Politechnica 34(3)255–265
Lőrincz J (1993a) On granular filters with the help of grading entropy. In: Conf on Filters in Geotechnical and Hydr Eng Brauns, Heibaum, Schuler, BALKEMA:67–69
Lőrincz J (1993b) On particle migration with the help of grading entropy. In: Proc Conf Filters in Geotechnical Hydr Eng Brauns, Heibaum, Schuler, BALKEMA:63–65
Lőrincz J, Imre E, Gálos M, Trang QP, Telekes G, Rajkai K, Fityus I (2005) Grading entropy variation due to soil crushing, Int J Geomechanics 5(4):311–320
Press WH, Flannery BP, Teukolsky SA, Wetterling WT (1986) Numerical Recipes. Cambridge Univ Press
Rajkai K (1993) A talajok vízgazdálkodási tulajdonságainak vizsgálati módszerei. In: Búzás I (ed) Talaj-és agrokémiai vizsg. INDA4321 Kiadó, Bp. 115–160
Várallyay Gy (1973) A talajok nedvességpotenciálja és új berendezés annak meghatározására az alacsony (atmoszféra alatti) tenziótartományban. Agrokémia és Talajtan 22:1–22
Van Genuchten MT (1980) A closed form equation for predicting the hydraulic conductivity of unsaturated soils, Soil Sci Soc Am J 44:892–898

Physical Modeling of SWCC for Granular Materials

Yvonne Lins[1], Yazhou Zou[2], and Tom Schanz[1]

[1] Laboratory of Soil Mechanics, Bauhaus-Universität Weimar, Coudraystr. 11C, 99423 Weimar, Germany
yvonne.lins@bauing.uni-weimar.de, tom.schanz@bauing.uni-weimar.de

[2] Institute of Soil mechanics and Foundation Engineering, University of the Federal Armed Forces, Munich, 85577 Neubiberg, Germany
yazhou.zou@unibw.de

Summary. A physically model was used in this study to estimate the soil–water characteristic curve (SWCC) of granular materials in case of Hostun sand. Based on soil parameters and pore geometry the main drainage and main imbibition process as well as a scanning drainage and scanning imbibition process was determined. In a modified pressure plate apparatus the SWCC was performed for different loading paths for a loose Hostun sand specimen. These experimental results were used to verify the physical model. Therefore, an inverse analysis was performed.

Key words: SWCC, physical modeling, sand, sensitivity analysis, parameter identification

1 Introduction

Properties of unsaturated soils are strongly related to the SWCC, also known as soil–water retention curve or capillary pressure–saturation relationship. Thus the SWCC is an important tool for determination of the hydraulic and mechanical behavior of unsaturated soils. The SWCC relates the volumetric water content, the gravimetric water content or the degree of saturation to the soil suction. Either the soil suction is increasing, which corresponds to the drainage process of the soil or the soil suction is decreasing, which corresponds to the imbibition process of the soil. Since the drainage and imbibition process are not unique but hysteretic in their behavior the engineers also have to deal with the so called scanning behavior. The scanning curves are located in between the main drainage curve and the main imbibition curve. Both the two main curves are called also boundary hysteresis curves.

To estimate the functional relationship of the SWCC empirical models and physical models are available. Physical models involve independent domain models as well as dependent domain models based on the domain theory (Néel

1942, 1943, Everett and Smith 1954, Mualem 1974, 1984) and models based on soil parameters (e.g. grain-size distribution, void ratio) and/or pore geometry (Aubertin et al. 2003). In Stoimenova et al. (2005) modelling of SWCC and the goodness of the fit are discussed. Empirical models (van Genuchten 1980, Fredlund and Xing 1994) use statistical analysis to best fit experimental data to the selected equation. However, in these empirical models parameters have no clear physical meaning, and to simulate the boundary curves of the same soil a model parameter can have different values. A procedure for modeling of scanning curves within one model is given in Stoimenova et al. (2006). Domain models use the pore–water distribution function to estimate the SWCC. To estimate the hysteretic scanning curves domain models are usually based on the knowledge of the boundary curves. Physical models utilize soil properties as well as geometric properties to estimate the SWCC.

Experimental determination of the SWCC is a time consuming process. Hence it is worthwhile to describe the relationship between volumetric water content based on soil properties and geometric properties. Zou (2003, 2004) developed a physical model, which is taking main imbibition process and main drainage process as well as drainage and imbibition processes located between the main drainage and imbibition loop into account. In this paper the physical model is further verified using experimental data.

2 Material Used

The material used in this study is Hostun Sand, a reference sand well studied in the research literature (Flavigny et al. 1990, Schanz 1998). According to the USCS classification the material is a poorly-graded medium sand SP.

3 Experimental Programm and Results

Test was conducted for loose ($e_0 = 0.89$) sand specimen in a modified pressure plate apparatus, which enables the determination of the SWCC for both drainage and imbibition processes (Lins and Schanz 2005). Overall volume changes of the specimens are measured by an attached dial gage.

The specimen ring is 71 mm in diameter and 20 mm high (Fig. 1). The ceramic disk used below the soil specimen and above the water reservoir has an air-entry value of 100 kPa. A burette with a capacity of 25 cm^3 is connected to the water reservoir at the bottom of the cell. During drainage and imbibition of the soil specimen water inflow and outflow was measured. A coarse porous stone is placed on the top of the soil specimen. Air pressure was applied to the top of the specimen through the coarse porous stone. It is well known that granular materials show a relatively small range of suctions over which the soil becomes unsaturated. Thus the cell uses hanging column technique when applying low suction values Haines (1930). By lowering the attached

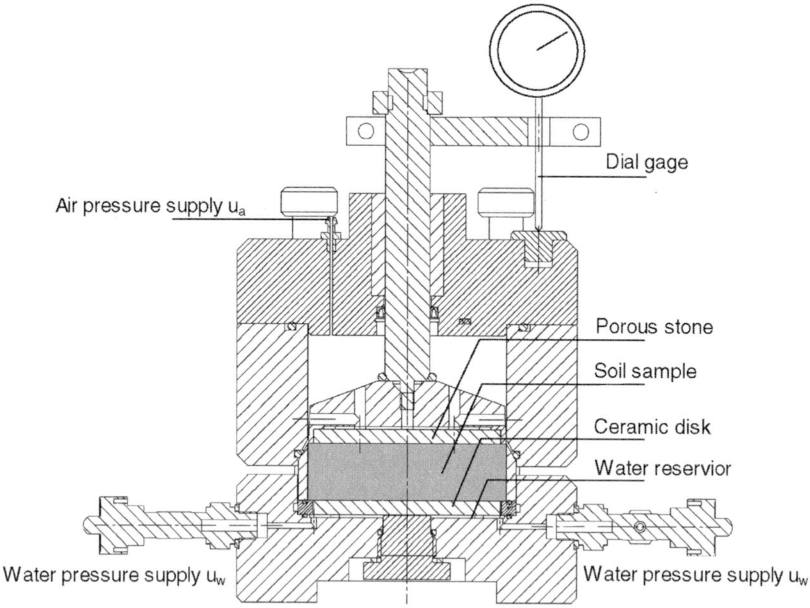

Fig. 1. Modified pressure plate apparatus

burette with respect to the top of the ceramic disk, suctions up to 4.0 kPa in steps of 0.1 kPa, could be applied to the specimen. For taking precise readings of water outflow and inflow the burette has a resolution of 0.05 cm^3. An air pressure system is applied to the top of the cell. By using the axis-translation technique pressure mode Hilf (1956) suctions up to 100 kPa could be applied to obtain test results in residual zone along the SWCC.

Dry sand specimen with a predetermined void ratio was prepared as a fixed ring specimen. By supplying water from the burette through the bottom ceramic disk the specimen was saturated. The initial specimen was water saturated specimen, which was loaded with predetermined suction increments using axis-translation technique, either suction mode test or pressure mode test. It has to be mentioned that the next suction increment was not applied before reaching equilibrium condition (i.e., no water inflow or outflow) in the specimen and thus the experiment is a steady-state type test.

The final gravimetric water content was calculated by oven-drying the specimen at the end of the test. Volumetric water contents, degree of saturations and gravimetric water contents corresponding to each suction step were back-calculated from these final values using incremental amounts of water flow from each step.

The experimental program consists of drying the specimen from saturated stage in several steps up to $\psi = 50.0$ kPa (initial drainage curve), wetting the specimen up to $\psi = 0.1$ kPa (main imbibition curve), drying the specimen up to $\psi = 2.6$ kPa (along the main drainage curve), wetting the curve up to

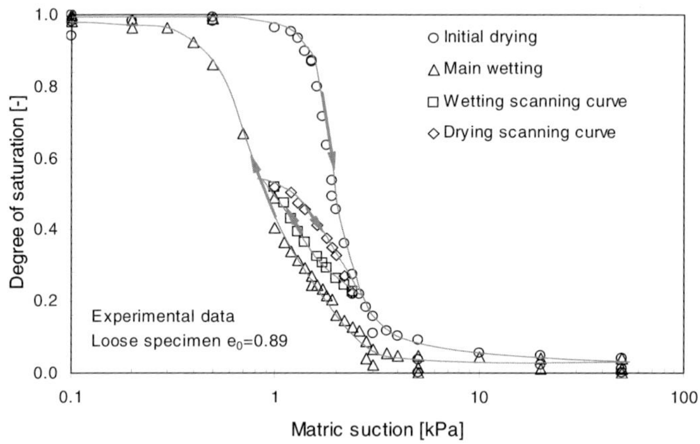

Fig. 2. Experimental results loose specimen including wetting and drying scanning curves

the main imbibition curve (scanning imbibition curve) as well as drying the specimen up to the main drainage curve (scanning drainage curve).

Experimental results for this loading path are given in Fig. 2. In the diagram the degree of saturation is plotted versus matric suction. The initial specimen is a fully fluid saturated specimen with $S_r = 1$. Already at a low suction value ($\psi = 1.0\,\text{kPa}$) the sand specimen starts to desaturate and the degree of saturation is decreasing. With increase in suction the degree of saturation is decreasing in a narrow range of low suction. At suction of $\psi = 5.0\,\text{kPa}$ nearly all pores are drained. The imbibition process shows that at a suction of approximately $\psi = 3.0\,\text{kPa}$ the specimen starts to adsorb water. The specimen is saturated at $\psi = 0.3\,\text{kPa}$. It can be seen from Fig. 2 that the drainage as well as imbibition curve are not unique but hysteretic in their behavior. The scanning drainage and scanning imbibition curves are located between the boundary hysteresis curve. At a given suction value the degree of saturation is for the drainage curve larger than for the imbibition curve.

4 Zou's Model for Determination of SWCC

To predict the boundary hysteresis curves and the scanning curves of unsaturated soils a physical model was proposed by Zou (2003, 2004). In this model, pores in an unsaturated soil are considered to be composed of the "contact regions" between two tangent soil particles and the irregular frustum-shaped pores among several adjacent particles (Fig. 3(**a**)). It is also assumed that all the particles in the soil are sphere-shaped and have the same sphere radius r_s and that all the frustum-shaped pores are regular symmetrical frustum-shaped pores which have the same height r_s, the same lower radius r_1 and the

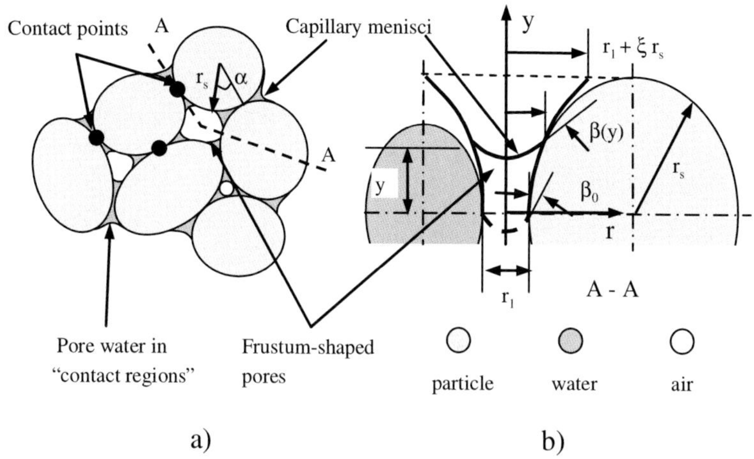

Fig. 3. Pore spaces and pore water

same upper radius $r_1 + \xi \cdot r_s$ (Fig. 3(**b**)). The average number of the contact points of a particle with its neighbours is termed as contact number n_c. The average number of the frustum-shaped pores per particle is called as a *frustum number* n_{f0}.

During a main wetting process of a soil, first all the "contact regions" are at the same time filled with water from the position angle $\alpha = 0$ to $\alpha = \alpha_{\max}$ (Fig. 3(**a**)). For $\alpha = \alpha_{\max}$ some adjacent capillary menisci contact with each other, and some frustum-shaped pores are formed. Then, all the frustum-shaped pores are at the same time filled with water from $y = 0$ to $y = r_s$ gradually (Fig. 3(**b**)). During a main drying process of a soil, first a part of the frustum-shaped pores begins to drain gradually from $y = r_s$ to $y = 0$. Then another part begins to drain and one after another. Subsequently the "contact regions" begin to drain from $\alpha = \alpha_{\max}$ to $\alpha = 0$, and a part of frustum-shaped pores with the number $\mu \cdot n_{f0}$ (where μ is a percentage) is just drained during the drying process of the "contact regions." If in a soil all the "contact regions" have been filled with water to $\alpha = \alpha_{\max}$ and all the frustum-shaped pores have been filled with water to $y = y_1 < r_s$ during a previous main wetting process, and now the soil begins to dry, then the drying process is called as secondary drying process. Similar to the main drying process, during the secondary drying process, first a part of the frustum-shaped pores that have been filled with water to $y = y_1$ previously begins to drain gradually. Then another part begins to drain and one after another. Subsequently the "contact regions" begin to drain from $\alpha = \alpha_{\max}$ to $\alpha = 0$, and a part of frustum-shaped pores with the number $\mu \cdot n_{f0}$ is also just drained during the drying process of the "contact regions." If in a soil only a part of the frustum-shaped pores has been drained to $y = y_2 < r_s$ during a previous main drying process, the other part are still filled with water to $y = r_s$, and now the soil begins to be wetted again, then the wetting process is called secondary wetting process. During the secondary wetting process only the part of the frustum-shaped pores that have been drained to $y = y_2$ previously are at the same time wetted again.

The volume $V_{w1}(\alpha)$ of the pore water in a "contact region" between two tangent particles (spheres) with the same radius r_s, depending on α, can be expressed approximately as

$$\frac{V_{w1}(\alpha)}{r_s^3} = 2\pi(1-\cos\alpha)\left[\sin^2(\alpha) - \frac{(1-\cos\alpha)(2+\cos\alpha)}{3}\right]. \tag{1}$$

The average pore volume V_{v0} per particle in a soil with pore ratio e, in proportion to r_s^3, can be written as

$$\frac{V_v}{r_s^3} = \frac{4\pi\cdot e}{3}. \tag{2}$$

When the "contact regions" in a soil are wetted or dried to a position angle α, according to the so-called capillary law (Fredlund and Rahardjo 1993) the dimensionless suction $\sigma_u(\alpha)$ in pore water, in relation to α can be expressed as follows

$$\sigma_u(\alpha) = \frac{(u_a - u_w)\cdot r_s}{T_s} = \frac{(2-\sin\alpha - 2\cos\alpha)\cdot\cos\alpha}{1-\sin\alpha(1-\cos\alpha)-\cos\alpha(2-\cos\alpha)} \tag{3}$$

where u_a and u_w are air and water pressure respectively, T_s is the so-called surface tension of capillary water (e.g. $T_s = 72.75\,\mathrm{mN/m}$ for the temperature $T = 20°\mathrm{C}$).

In Zou (2003, 2004) it was proposed that the form of the ideal symmetrical frustum-shaped pores can be described using a function y relating to the radius r (Fig. 3(**b**)) as follows

$$\frac{y}{r_s} = a - \frac{b}{r^2/r_s^2 - c} \qquad (0 \le y \le r_s) \tag{4}$$

where a, b and c are three constants that can be determined according to geometrical and physical boundary conditions using following equations

$$a(a-1)\left(\ln\frac{a}{a-1} - \frac{1}{a}\right)\left[\rho((\alpha_{\max}) + \xi)^2 - \varrho^2(\alpha_{\max})\right] = \frac{4e}{3n_{f0}} - \frac{n_c\cdot V_{w1}(\alpha_{\max})}{2\pi\cdot n_{f0}\cdot r_s^3} - \rho^2(\alpha_{\max}) \tag{5}$$

$$b = a(a-1)[(\rho(\alpha_{\max}) + \xi)^2 - \rho^2(\alpha_{\max})] \tag{6}$$

and

$$c = \rho^2(\alpha_{\max}) - \frac{b}{a} \tag{7}$$

where $\rho(\alpha_{\max}) = (1-\cos\alpha_{\max})/\cos\alpha_{\max}$ and ξ is a form parameter which describes the form of the symmetrical frustum-shaped pores.

When a symmetrical frustum-shaped pore is filled with water up to y, from Eq. (4) and by integration the volume $V_{w1}(y)$ of the pore water in the symmetrical frustum-shaped pore can be written as

$$\frac{V_{w1}(y)}{r_s^3} = \pi\left(c\frac{y}{r_s} + b \cdot \ln\frac{a}{a - y/r_s}\right) \qquad (0 \le y \le r_s)\,. \tag{8}$$

The total volume $V_w(y, \alpha)$ of pore water per particle during the main wetting and drying processes as well as during the secondary wetting and drying processes, depending on y and/or α, can be written as

$$\frac{V_w(y,\alpha)}{r_s^3} = n_f(y)\frac{V_{w1}(y)}{r_s^3} + n_{fi}(y,\alpha)\frac{V_{w1}(y_i)}{r_s^3} + n_c\frac{V_{w1}(\alpha)}{2r_s^3} \tag{9}$$
$$(0 \le y \le r_s;\ 0 \le \alpha \le \alpha_{\max},\ i = 0, 1, 2)$$

where $n_f(y)$ is the average number of the frustum-shaped pores that are wetting or drying per particle, $n_{fi}(y, \alpha)$ is the average number of the frustum-shaped pores that are filled with water to $y = y_i$ (where y_i is the maximal filling height). $n_f(y)$, $n_{fi}(y)$, y_i and the position angle α in Eq. (9) can be determined corresponding to Table 1 for different wetting and drying processes.

The degree of saturation $S_r(y, \alpha)$ depending on y and/or α can be estimated using the following equation

$$S_r(y,\alpha) = S_{r0}\frac{V_w(y,\alpha)}{V_v} \tag{10}$$

where S_{r0} is the degree of saturation for suction $(u_a - u_w) \approx 0$.

To determine the suction in pore water during wetting and drying the frustum-shaped pores, in Zou (2003, 2004) it is assumed that the contact

Table 1. Model parameters referring to different loading paths

Processes	y_i	$n_f(y)$	$n_{fi}(y,\alpha)$	α
Main wetting				
"contact regions"	—	0	0	$0 \sim \alpha_{\max}$
frustum-shaped pores	$y_0 = r_s$	n_{f0}	0	$\alpha_{\max}$
Main drying				
frustum-shaped pores	$y_0 = r_s$	$n_{f0}(1-\mu)(1-y/r_s)$	$n_{f0}[\mu + (1-\mu)y/r_s]$	$\alpha_{\max}$
"contact regions"	$y_0 = r_s$	0	$\mu \cdot n_{f0} \cdot \alpha/\alpha_{\max}$	$0 \sim \alpha_{\max}$
Secondary drying				
frustum-shaped pores	y_1	$n_{f0}(1-\mu)(1-y/y_1)$	$n_{f0}[\mu + (1-\mu)y/y_1]$	$\alpha_{\max}$
"contact regions"	y_1	0	$\mu \cdot n_{f0} \cdot \alpha/\alpha_{\max}$	$0 \sim \alpha_{\max}$
Secondary wetting				
frustum-shaped pores	y_2	$n_{f0}(1-\mu)(1-y_2/r_s)$	$n_{f0}[\mu + (1-\mu)y_2/r_s]$	$\alpha_{\max}$

angle $\beta(y)$ (Fig. 3(**b**)) as a function of y can be described using the following equation

$$\beta(y) = \sqrt{\frac{B}{A - \frac{y}{r_s}} + C}\,. \tag{11}$$

The three constants A, B and C can be determined according to physical boundary conditions using the following equations

$$A = \zeta, \quad B = \zeta(\zeta - 1)\left(\frac{\pi^2}{4} - \beta_0^2\right), \quad C = \zeta \cdot \beta_0^2 - (\zeta - 1)\frac{\pi^2}{4} \tag{12}$$

where

$$\beta_0 = \arccos\left[\frac{1 - \sin\alpha_{\max}/2 - \cos\alpha_{\max}(1 - \cos\alpha_{\max})}{1 - \sin\alpha_{\max}(1 - \cos\alpha_{\max}) - \cos\alpha_{\max}(2 - \cos\alpha_{\max})}\right] \tag{13}$$

and $\zeta > 1$ is a meniscus parameter which describes the physical properties of soil particle surfaces.

According to the so-called capillary law (Fredlund and Rahardjo 1993), the dimensionless suction $\sigma_u(y)$ in pore water during wetting and drying the frustum-shaped pores, depending on r and β and so that on y, can be expressed as

$$\sigma_u(y) = \frac{(u_a - u_w) \cdot r_s}{T_s} = \frac{2\cos\beta}{r/r_s} = \frac{2\cos \cdot \left(\frac{B}{A - y/r_s} + C\right)^{0.5}}{\left(\frac{b}{a - y/r_s} + c\right)^{0.5}}\,. \tag{14}$$

Now using equations (10) and (3) the relationship between the degree of saturation $S_r(\alpha)$ and the suction $\sigma_u(\alpha)$ during wetting and drying the "contact regions" for $0 < \alpha \leq \alpha_{\max}$ can be calculated, and using equations (10) and (14) the relationship between $S_r(y)$ and $\sigma_u(y)$ during wetting and drying the frustum-shaped pores for $0 < y \leq r_s$ can be also estimated.

4.1 Sensitivity of the Parameters on the Shape of the Curve

The model includes seven parameters. To determine the influence of the parameters on the shape of the curves an analysis was performed. While the influence of the parameter n_c is negligible the parameters $\alpha_{\max}$, n_{f0}, ξ, ζ, μ and S_{r0} are influencing the shape of the SWCC. To clarify the influence of the parameters on the shape of the SWCC, it is classified into three sections, namely the saturated zone, the unsaturated zone as well as the residual zone. Additionally typical parameters for the air-entry value, ψ_{aev}, the saturated volumetric water content, θ_s, and the residual volumetric water content, θ_r, with the corresponding residual suction, ψ_r, and the water-entry value, ψ_{wev} are included in Fig. 4. The parameter ξ is responsible for the geometry of the frustums, the parameter n_{f0} describes the number of the frustums, and the parameter μ describes the drainage process of some residual frustums, thus they are responsible for the volume of water in the frustum-shaped pores, and influence mainly the saturated and the unsaturated zone.

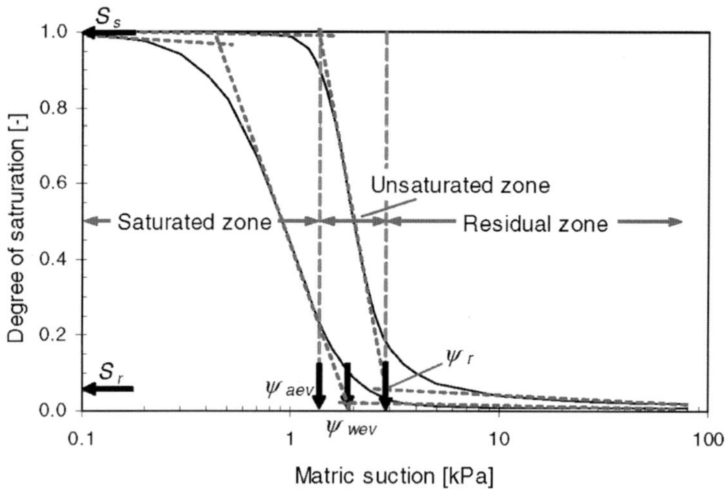

Fig. 4. Typical parameters and zones of the SWCC

Maximum contact angle α_{max}

The maximum contact angle α_{max} is reached, when adjoining contractile skins are in contact and the frustum-shaped pores are filled with water. As smaller the contact angle α_{max} as higher the suction, which has to be applied for drying the contact regions. With decreasing contact angle α_{max} the unsaturated zone is increasing and the residual zone is decreasing during drainage process (see Fig. 5).

Frustum number n_{f0} per sphere

When α_{max} is reached, the frustum-shaped pore between the grains is filled with water. The remaining pore volume is taken into account by a function, describing the volume of a frustum. The largest volume of water is located in these pores. Thus, the number of frustums n_{f0} is influencing the shape of the curve significantly in the saturated zone as well as unsaturated zone. With increasing n_{f0} the saturated zone and unsaturated zone is shifting to higher suction values during drainage process. Same behavior can be observed for the imbibition process. As larger n_{f0} as larger the air-entry value ψ_{aev} and the water-entry value ψ_{wev} (see Fig. 6).

Form parameter ξ

The form parameter ξ defines the geometry of the frustum and estimates the upper diameter of the frustum. With increasing ξ the curve becomes flatter

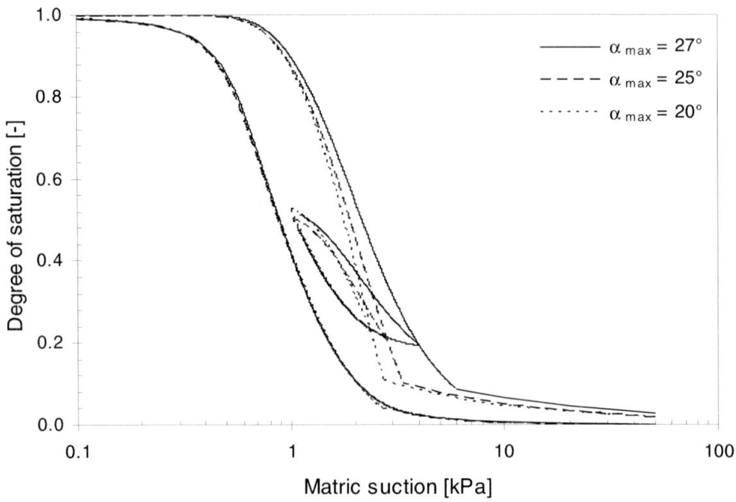

Fig. 5. Influence of the parameter $\alpha_{\max}$ on the shape of the SWCC

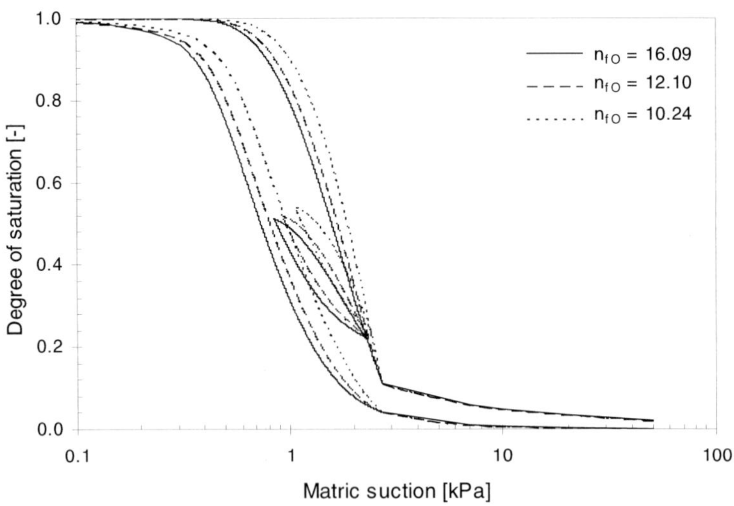

Fig. 6. Influence of the parameter n_{f0} on the shape of the SWCC

for drainage and imbibition process. As smaller ξ as more distinct the air-entry value ψ_{aev}. The saturated zone is shifting to higher suction values for decreasing ξ while wetting the specimen (see Fig. 7).

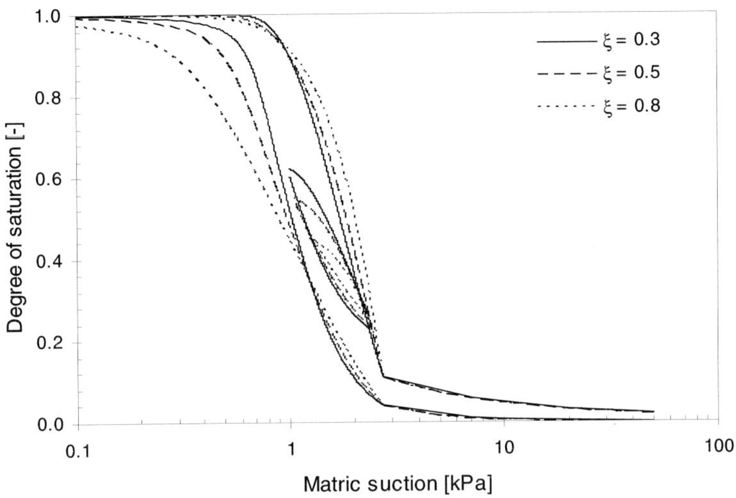

Fig. 7. Influence of the parameter ξ on the shape of the SWCC

Material parameter ζ

The material parameter ζ is responsible for the form of the meniscus in the frustum-shaped pores and thus it influences the suction $(u_a - u_w)$ when the frustum-shaped pores are wetting or drying. Furthermore, it influences the saturated and unsaturated zone. The larger ζ is, the larger the air-entry value ψ_{aev}.

Fraction of frustums μ

The fraction of frustums μ only influences the drainage process. During the drainage process the larger frustum-shaped pores are drying first while the smaller pores still remain water. With increasing μ the residual degree of saturation S_r is increasing for the corresponding residual suction ψ_r (see Fig. 8).

Degree of saturation S_{r0} for $(u_a - u_w) \approx 0$

During the wetting process of a soil some air bubbles can be closed in pore water. If the ratio of the net volume of pore water to the total volume of pore water (including the volume of the closed air bubbles) during the wetting process is assumed to be constant, then the ratio is just the degree of saturation S_{r0} for $(u_a - u_w) \approx 0$. The parameter S_{r0} influences mainly the saturated zone and thus influences the air-entry value ψ_{aev}.

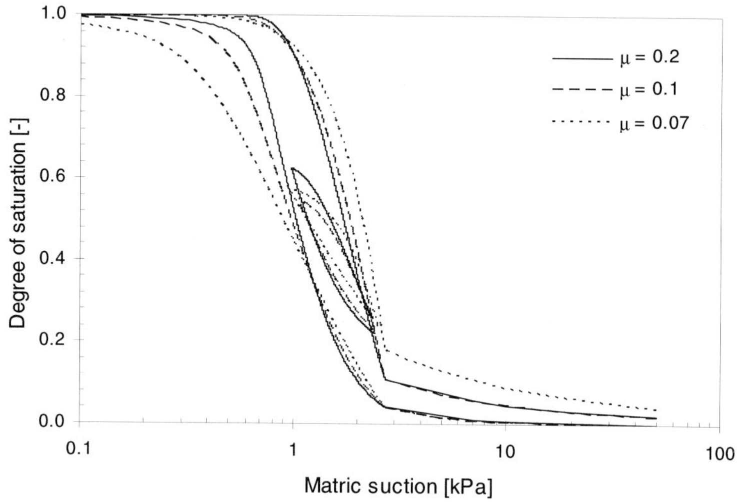

Fig. 8. Influence of the parameter μ on the shape of the SWCC

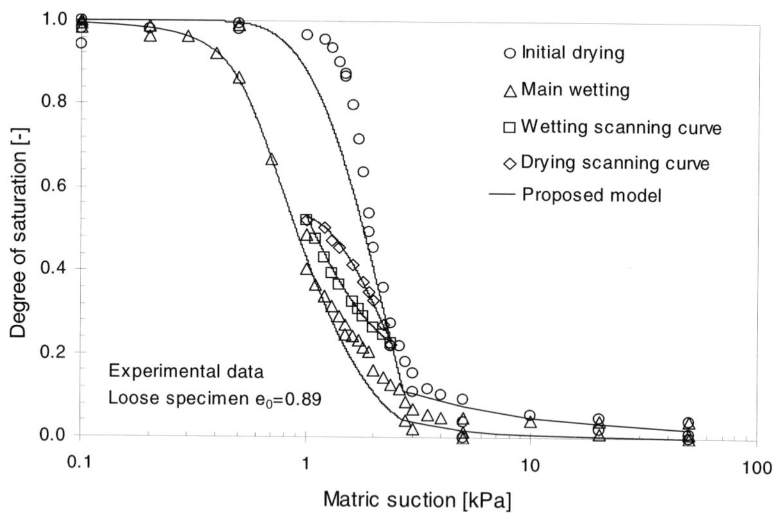

Fig. 9. Curve fitting result loose specimen

4.2 Verification of the Model Using Experimental Results

The best fit of the experimental results is given in Fig. 9, where the degree of saturation is plotted versus matric suction. The best fit was found by using an inverse analysis, where experimental and calculated results were compared with an optimization routine (VARO^{2}PT) Zimmerer and Lobers (2005). In a first step the parameters are set by conventional means followed by a first simulation. Based on the estimated parameter set the difference between experimental and calculated results are determined. Then the objective function of the optimization routine is modified until the difference between the experimental and calculated results is minimized, which indicates the best fit (Zimmerer and Schanz 2006, Schanz et al. 2006).

During drainage process the experimental results are in good agreement with the curve fit in the saturated zone and in the residual zone. Differences between experimental and calculated data can be observed in the unsaturated zone. During imbibition process the experimental results are close to the calculated results. For both the scanning drainage and scanning imbibition process a close fit was found.

5 Conclusion

A physically based model was used in this study to estimate the SWCC of Hostun sand. The model was verified by using experimental results from a modified pressure plate apparatus of a loose Hostun sand specimen.

Good agreement between the experimental and the theoretically calculated results was found for the boundary curves as well as the including scanning drainage and scanning imbibition curves. During main drainage curve larger differences between experimental results and calculated results were found around the air-entry value ψ_{aev}. Since the air-entry value ψ_{aev} is a sensitive part of the SWCC, it is suggested to improve the model at this point by taking into account the suction value, when pores start to drain.

Nevertheless the advantage of the model is, that with soil mechanical and geometrical parameters only, it enables the estimation of the SWCC based on physical understanding.

Acknowledgements

The present study was carried out in the frame of the DFG-research project *Mechanik teilgesättigter Böden* (Mechanics of unsaturated soils). An acknowledgement is given to the research assistantship provided by the Bauhaus-Universität Weimar, Germany, through the research grant from Deutsche Forschungsgemeinschaft (DFG), FOR 444/2.

References

Aubertin M, Mbonimpa M, Bussiere B, Chapius RP (2003) A model to predict the water retention curve from basic geotechnical properties, Can Geotech J 40:1104–1122

Everett DH, Smith FW (1954) A general approach to hysteresis: 2. development of the domain theory, Trans Faraday Society 50:187–197
Flavigny E, Desrues J, Palayer B (1990) Note technique: Le sable d'Hostun RF, Rev Franc Géotech Rev Franc Géotech 53:67–70
Fredlund DG, Rahardjo H (1993) Soil mechanics for unsaturated soils. John Wily and Sons Inc., New York
Fredlund DG, Xing A (1994) Equations for soil-water characteristic curve, Can Geotech J 31:521–532
Haines WB (1930) The hysteresis effect in capillary properties and the modes of moisture distribution associated therewith, J Agr Sci 20:96–105
Hilf JW (1956) An investigation of pore-water pressure in compacted cohesive soils. US Dept of the Interior, Bureau of Reclamation, Design and Const Div, Denver Tech Memo 654
Lins Y, Schanz T (2005) Determination of hydro-mechanical properties of sand. In: Schanz T (ed) Unsaturated Soils: Experimental Studies, Springer-Verlag, Berlin Heidelberg, 15–32
Mualem Y (1974) A conceptual model of hysteresis, Water Resources Research 10:514–520
Mualem Y (1984) A model dependent domain theory of hysteresis, Soil Science 137:283–289
Néel L (1942) Théories des lois d'aimantation de Lord Raileigh, Cah Phys 12:1–20
Néel L (1943) Théories des lois d'aimantation de Lord Raileigh, Cah Phys 13:19–30
Schanz T (1998) Zur Modellierung des mechanischen Verhatltens von Reibungsmaterialien, Institut für Geotechnik, Universität Stuttgart, Mitteilung 45
Schanz T, Zimmerer MM, Datcheva M, Meier J (2006) Identification of constitutive parameters for numerical models via inverse approach, Felsbau 24:11–21
Stoimenova E, Datcheva M, Schanz T (2005) Statistical approach in soil–water characteristic modelling. In: Springer Proceedings in Physics 93/II:189–200
Stoimenova E, Lins Y, Datcheva M, Schanz T (2006) Inverse modelling of soil hydraulic characteristic functions, In: Gürlebeck K, Könke C (eds) 17th Int Conf on the Application of Computer and Mathematics in Architecture and Civil Engineering, Weimar, Germany, 12–14 July 2006
van Genuchten MTh (1980) A closed-form equation for predicting the hydraulic conductivity of unsaturated soils, Soil Sci Soc 44:892–898
Zimmerer MM, Lobers S (2005) VARO^{2}PT-Software for inverse analysis, VAROCON-Software and engineering for geotechnical application Weimar, `www.varocon.com`
Zimmerer MM, Schanz T (2006) Determination of soil parameters for modeling of deep excavations utilizing an inverse approach. In: Triantafyllidis (eds) Numerical modelling of construction processes in geotechnical engineering for urban environment, Taylor and Francis Group, London, 21–28
Zou Y (2003) Ein physikalisches Modell der pF-Kurve für teilgesättigte grobkörnige Böden, Bautechnik 80:913–921
Zou Y (2004) Ein physikalisches Modell der pF-Kurve für teilgesättigte grobkörnige Böden auf primäre Entwässerung und sekundäre Be- und Entwässerung, Bautechnik 81:371–378

Implications of a Generalized Effective Stress on the Constitutive Modelling of Unsaturated Soils

Mathieu Nuth and Lyesse Laloui

Soil Mechanics Laboratory, Ecole Polytechnique Fédérale de Lausanne (EPFL), Station 18, 1015 Lausanne, Switzerland
mathieu.nuth@epfl.ch, lyesse.laloui@epfl.ch

Summary. The major consequences of the application of the effective stress concept to unsaturated soils are investigated. Several experimental data sets are re-examined in the light of a generalized effective stress, intended to describing the mechanical behaviour of porous media. The critical state lines at different saturation states tend to converge remarkably toward a unique saturated line in the deviatoric stress versus mean effective stress plane. The effective stress interpretation is also applied to isotropic paths and compared with traditional net stress conception. The accent is finally laid on the key feature for constitutive frameworks based on a unified stress, namely the sufficiency of a unique mechanical yield surface.

Key words: effective stress, unsaturated soils, constitutive modelling, critical state

1 Introduction

Even though the long-time debate on stress frameworks for constitutive modelling of unsaturated soils has eventually come to an end, evidencing the need for a combination of hydro-mechanical stresses, the forms of possible effective stresses are still discussed. We thus study the implications of the use of the Bishop's generalised effective stress, to be defined later, for the interpretation of the unsaturated mechanical behaviour. The hereby assertions are exclusively based on experimental data and could consist the pretext of an advanced constitutive model. The principle is to re-plot experimental data sets expressed in terms of classical unsaturated stress variables, which are net stress and suction (Fredlund and Morgenstern 1977), into new effective stress planes. The uniqueness of critical state line for different suctions, and modifications along unsaturated mechanical loading paths are investigated, leading to introductory discussion on modelling.

2 Unified Stress Framework for Unsaturated Soils

An essential preliminary to constitutive modelling of multi-phase porous media, such as unsaturated soils, is the determination of the effective stress of the solid skeleton. The effective stress converts a multi-phase porous medium to a mechanically equivalent, single-phase, single-stress state continuum. Consequently, the constitutive equations for the mechanical behaviour directly link the change in strain to a variation in a single stress that is averaged over a volume comprehending several constituents, each of which being likely to react internally to a global external load.

On this basis, a general definition of the effective stress σ'_{ij} in a porous medium saturated with n fluids is the following:

$$\sigma'_{ij} = \sigma_{ij} + \sum_{k=1}^{n} \alpha_k u_k \delta_{ij} \tag{1}$$

where σ_{ij} is the exterior stress, δ_{ij} the Kroenecker's delta, u_k the pressure of fluid phase k, and α_k the corresponding scaling factor. Terzaghi's expression for saturated soils (Terzaghi 1943) enters this family of effective stresses; so does Bishop's (1959) approach for unsaturated media:

$$\sigma'_{ij} = (\sigma_{ij} - u_a\delta_{ij}) + \chi(u_a - u_w)\delta_{ij} = \sigma_{\mathrm{net}ij} + \chi s \tag{2}$$

with u_a being the pore air pressure, u_w the pore water pressure, $\sigma_{\mathrm{net}ij}$ the net stress and s the matric suction. χ is the effective stress parameter, varying from 0 to 1 to cover respectively the range of fully dry to fully saturated states. A possible choice is written (Schreffler 1984):

$$\chi = S_r = \frac{\text{volume of water}}{\text{volume of voids}} \tag{3}$$

The previous identity (3) along with Eq. (2) give the definition of the Bishop's generalised effective stress, also called average skeleton stress (Jommi 2000). From a constitutive point of view and under given assumptions, thermodynamic (Hutter et al. 1999), as well as energetic (Houlsby 1997) approaches identify this appropriate form of the effective stress which must be associated with a work-conjugate strain variable. In parallel, the evolution of the degree of saturation is necessary to fully describe the behaviour of the unsaturated soil. It is conjugate to suction scaled by porosity. Laloui and Nuth (2005) provide a deepened discussion on the latter stress framework. It is proposed hereafter to show that no complexity is brought along by the adoption of such a framework compared with the two independent stress variables approach, using net stress and suction.

3 Critical State Analysis

In the following, the conventional triaxial stresses are used for representations. The deviator stress q and mean stress p are defined as follows:

$$q = \hat{\sigma}_{11} - \hat{\sigma}_{33} \tag{4}$$

$$p = \frac{\hat{\sigma}_{11} + \hat{\sigma}_{22} + \hat{\sigma}_{33}}{3} \tag{5}$$

$\hat{\sigma}_{ij}$ being the stress state variable chosen, depending on framework.

Khalili et al. (2004) investigated successfully the uniqueness of the critical state line (CSL) in a deviator stress versus effective mean stress plane for different levels of suction. The stress variable used is a Bishop-type effective stress (Eq. (2)) with a particular χ. We propose to reinterpret similarly several shearing datasets with the Bishop's generalised effective stress, termed 'effective stress' in the following for the sake of simplicity.

Figures 1 and 2 mirror two stress interpretations for CSL, namely the $(q - p_{\text{net}})$ plane (Figs. 1**a** and 2**a**) and$(q - p')$ plane (Figs. 1**b** and 2**b**). Most of

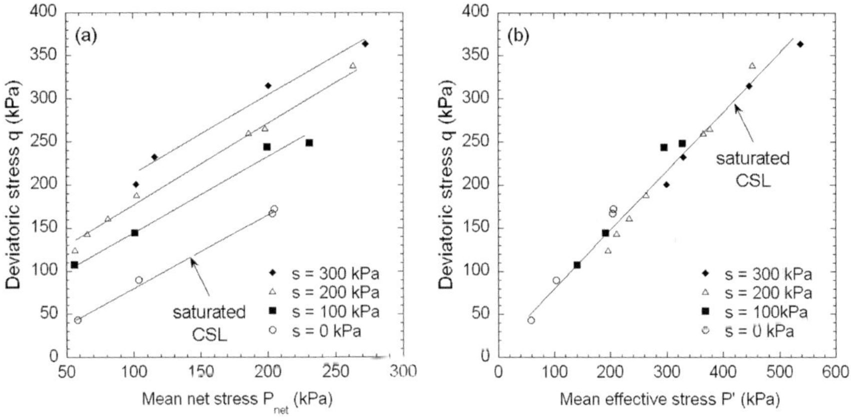

Fig. 1. Critical state lines for Kaolin at different suctions (Sivakumar 1993)

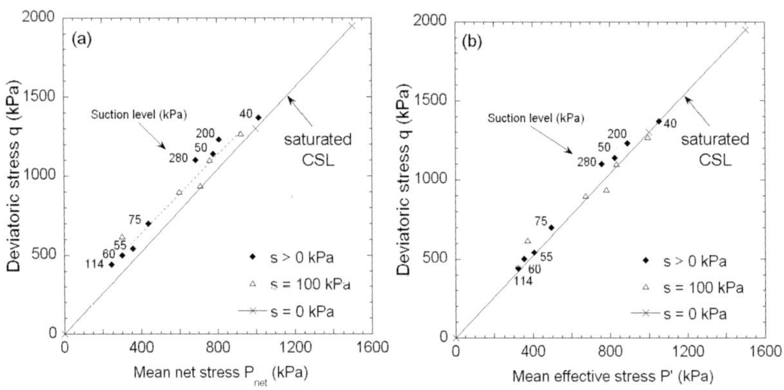

Fig. 2. CSL for Sion silt at different suctions. Data from Geiser et al. (2006)

unsaturated experimental behavioural studies report data in traditional net stress planes, the latter being an experimentally controlled stress variable. Under this form, even though the variations of the friction angle and cohesion are evidently different from one material to another, Figs. 1**a** and 2**a** highlight a plurality of critical state lines in the net stress planes for both materials.

Whereas transforming the plot from net stress conception to effective stress representation does not affect deviatoric stress level, the isotropic effect of suction scaled by degree of saturation is added to the net mean stress. The critical state lines at different suctions are uniformly translated horizontally by the amount of stress mentioned above, the immediate consequence of which is a reduction of apparent cohesion. The alignment property of points is conserved in the new plane as well as friction angles.

The processed results of Sivakumar (1993) in Fig. 1**b** are particularly accounting for an obvious unification of the effective critical state whatever the level of suction between 0 and 300 kPa. Concerning Sion silt (Geiser et al. 2006), while the net stress interpretation (Fig. 2**a**) already evidences a narrow arrangement of lines, the effective stress version (Fig. 2**b**) tends to align even more the experimental points, and sets the whole scattering much closer to the saturated CSL, accounting for its uniqueness.

This encourages simplification of parameter determination, assuming that saturated shear parameters are sufficient to describe both saturated and partially saturated critical state behaviours, overcoming the difficulties linked to the suction-dependent cohesion observed in the $(q - p_{\text{net}})$ plane.

4 Unsaturated Mechanical Compression

Detractive arguments regarding an effective stress framework concern the apparent difficulty to evaluate the stress level and then to interpret results in subsequent effective planes. However, even if the stress variables are not directly the experimentally controlled ones, that are net stress and suction, the elected generalised effective stress is a rather simple combination of them, with the parallel retention information. Nevertheless, data sets lacking from continuous hydraulic monitoring (that is knowledge of water content, volumetric strain and matric suction) are obviously excluded from the proposed effective stress interpretation.

Figure 3 indicates that the increase in compressibility and preconsolidation pressure with suction, observed in the experimental net stress mechanical planes (Fig. 3**a**) are recovered in the effective stress interpretation (Fig. 3**b**). On the whole, those two suction effects appear to be amplified with the proposed generalised effective stress approach.

Gallipoli et al. (2003) plotted a similar interpretation with the Bishop's generalised effective stress, interpreting slightly curved shapes for compressions rather than lines. However, Fig. 3 shows that a linear model could fit

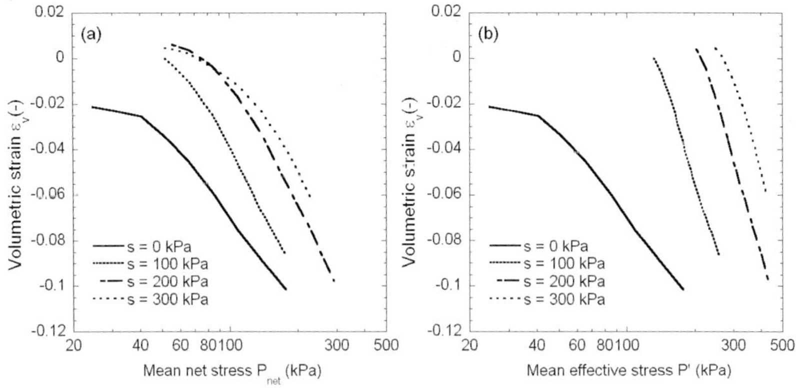

Fig. 3. Isotropic compressions of kaolin at different suctions (Sivakumar 1993)

well the compression lines in the plastic part; and that inflexions of the experimental curves, if some, are likely to appear only at high mean effective stresses or might not be significant.

Figure 3 also enables the determination of the Loading Collapse (LC) yield curve, which original definition is due to Alonso et al. (1990) in the matric suction versus mean net stress plane. This yield limit accounting for the increase in preconsolidation pressure p_c with suction is a reversible function that reflects the suction hardening, see Fig. 4. As declared previously, the shape of the LC curve in the effective stress interpretation is similar to the reference one, but neatly amplified. A significant consequence is that analogous mathematical formulations for the function $p_c = \hat{p}_c(s)$ could be applied to both net and effective stress analyses.

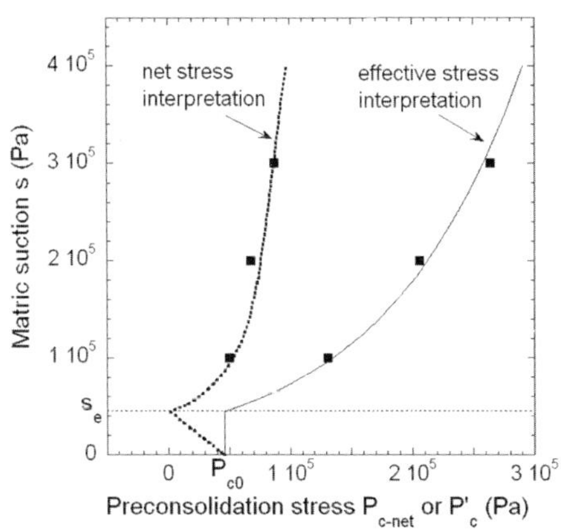

Fig. 4. Shape of the LC curve in two interpretations. Data from Sivakumar (1993)

Concerning the saturated domain, defined by $s < s_e$ with s_e being the air entry suction, the shape of the yield limit is a simple vertical line in the $(s - p'_c)$ plane. The effective stress interpretation induces this independency of the apparent preconsolidation pressure on suction, unlike in the $(s - p_{c-\text{net}})$ plane. Laloui and Nuth (2005) detail the implications of this bottom shape on volumetric strain upon hydro-mechanical loads.

5 Constitutive Modelling Framework

Statements in Section 2 imply that the effective stress is simply defined as that inducing the mechanical (elastic) strain of the solid skeleton. The constitutive equation for the mechanical behaviour is thus written:

$$\varepsilon^e_{ij} = C^e_{ijkl}\sigma'_{kl}. \tag{6}$$

Judging on Eq. (2), the effective stress is incremented by the application of any combination of the external mechanical stress σ_{ij} and the matric suction s. So, any of the two separate or combined loads induce variations in total volumetric deformation $\dot{\varepsilon}_v$ that is equivalent to a pure mechanical straining of the material, as expected from constitutive Eq. (6).

On this basis, the mechanical stress state being fully described by the means of the unified stress, plasticity mobilisation can be evaluated via the generalised stress state alone. A relevant consequence of the unified stress approach lies thus in the sufficiency of a single mechanical yield surface to comprehend any elasto-plastic behaviour resulting in a modification of the skeleton deformations. So, for the mechanical stress-strain behaviour, no second yield surface in suction is required, setting the model formulation free from any kind of Suction Increase (SI) or Suction Decrease (SD) yield curves, that are proper to constitutive models written in terms of independent stress variables such as Barcelona Basic Model family (Alonso et al. 1990) or (Laloui et al. 2001).

In the end, the Loading Collapse curve should include a hydro-mechanical coupling, which leads to the following formulation:

$$p'_c = \hat{p}'_c(\xi, s) \text{ for all } s \tag{7}$$

where the term ξ gathers the saturated hardening variables. It can be remarked that the expression for LC yield curve is continuous from the saturated state to the unsaturated state.

Visualizing the stress paths in the matric suction versus mean stresses planes help understanding better the unified stress framework implications (Fig. 5). Whereas a drying path under constant net stress is a simple vertical line in the $(s - p_{\text{net}})$ plane, the effective stress formulation confers a curved shape in the $(s - p')$ plane. By reference to Fig. 4, the effective drying path

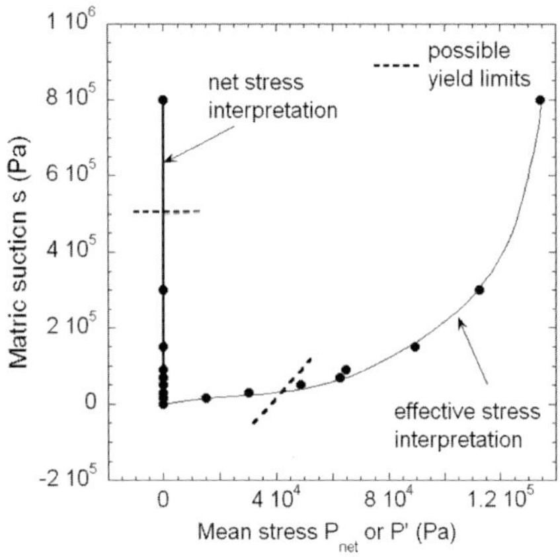

Fig. 5. Drying path in two interpretations on Sion Silt (Geiser et al. 2006)

evidences a possible direct yielding on the LC curve, whereas the net stress interpretation shows the need for a supplementary pseudo-horizontal yield limit. However, the curvature of the drying path in the effective stress conception is narrowly linked to the soil water retention curve shape, and some inconsistencies may appear at very high suctions, where the degree of saturation drops close to zero (Eq. (2)). A particular attention must be paid for interpreting this limit state.

6 Conclusion

The major implications of the generalised effective stress on the constitutive modelling of partially saturated media have been overviewed. On the whole, the use of a single effective stress for describing the mechanical behaviour tends to unify the critical state analysis and yielding features, favouring the adaptation of saturated models to unsaturated states.

References

Alonso EE, Gens A, Josa A (1990) A Constitutive Model for Partially Saturated Soils. Geotechnique 40(3):405–430

Bishop AW (1959) The principle of effective stress. Tecnisk Ukeblad 39

Fredlund DG, Morgenstern NR (1977) Stress variables for unsaturated soils. Journal of the geotechnical engineering division, ASCE 103(GT5):447–466

Gallipoli D, Gens A, Sharma R, Vaunat J (2003) An elasto-plastic model for unsaturated soil incorporating the effects of suction and degree of saturation on mechanical behaviour. Géotechnique, 53(1):123–135

Geiser F, Laloui L, Vulliet L (2006) Elastoplasticity of unsaturated soils: laboratory tests results on a remoulded silt. Soils and Fundations, 46(5).

Houlsby GT (1997) The work input to an unsaturated granular material. Géotechnique, 47(1):193–196

Hutter K, Laloui L, Vulliet L. (1999) Thermodynamically based mixture models of saturated and unsaturated soils. Journal of Mechanics of Cohesive-Frictional Materials 4:295–338

Jommi C (2000) Remarks on the constitutive modelling of unsaturated soils. Exp. Evidence & Theoretical Approaches in Unsaturated Soils, Trento, pp 139–153

Khalili N, Geiser F, Blight GE (2004) Effective stress in unsaturated soils: review with new evidence. International Journal of Geomechanics 4(2)

Laloui L, Geiser,F, Vulliet L (2001) Constitutive modelling of unsaturated soils. Revue Française de Génie Civil 5:797–887

Laloui L, Nuth M (2005) An introduction to the constitutive modelling of unsaturated soils. Revue Européenne de Génie Civil 9(5–6):651–669

Schreffler B (1984) The finite element method in soil consolidation (with applications to surface subsidence). Ph.D. Thesis. University College of Swansea

Sivakumar V (1993) A critical state framework for unsaturated soil. Ph.D. Thesis, University of Sheffield, Sheffield, UK

Terzaghi K (1943) Theoretical soil mechanics. Chapman and Hall, London

Shear Strength of Unsaturated Soils: Experiments, DEM Simulations, and Micromechanical Analysis

Vincent Richefeu, Moulay Saïd El Youssoufi, and Farhang Radjaï

LMGC UMR CNRS 5508, Université Montpellier 2, Place Eugène Bataillon
34095 Monpellier Cedex 5, France richefeu@lmgc.univ-montp2.fr,
elyous@lmgc.univ-montp2.fr, radjai@lmgc.univ-montp2.fr

Summary. We investigate shear strength properties of wet granular materials as a function of water content in the pendular state. Sand and glass beads were wetted and tested in a direct shear cell. In parallel, we carried out molecular dynamics simulations by using an explicit expression of capillary force as a function of interparticle distance, water bridge volume and surface tension. Experiments and numerical simulations are in good agreement. We show that the shear strength is mostly controlled by the distribution of liquid bonds. This property results leads to the saturation of shear strength as a function of water content. We arrive at the same conclusion by analyzing the shear strength from the microstructure and by accounting for particle polydispersity. Finally, we discuss the potentialities of the discrete element approach as applied to unsaturated soils.

Key words: granular materials, capillary cohesion, shear strength, DEM simulations, micromechanics

1 Introduction

The interest of scientists for granular materials is motivated by the important place they occupy in our natural environment. Well-known examples are soils, rocks and powders. In addition, many industrial processes implement transformations of granular materials such as pharmaceutical tablets, agricultural and food products, building materials (Bika et al. 2001, Iveson et al. 2002). The mechanical behavior of these materials, in particular soils, was widely developed in the context of macroscopic continuum approach. This approach is based on the concept of Representative Elementary Volume (REV) which is not well-established for granular media. Moreover, it introduces many parameters that require a large number of identification experiments. In addition, most continuum approaches do not account for grain-scale mechanisms such as the initiation and propagation of shear bands or cracks.

On the other hand, the discrete character of granular materials makes it possible to identify unambiguously the grain scale and thus the proper kinematics and mechanical behavior at this scale. Grain-scale properties depend on composition, form, size or surface state of grains. A discrete approach requires furthermore a rigorous definition of the interactions of each grain with its close neighbors through mechanical contacts, friction and/or cohesion. External actions like the moisture, temperature and pressure can also act on the evolution of these interactions. This is a promising approach for understanding all the complexity and often surprising properties of granular materials if adequate techniques can be developed. Such techniques are often referred to as Discrete Elements Method (DEM). This method has been successfully applied to study classical problems in soil mechanics as shear bands localization (Bardet and Proubet 1991), slope stability (Staron et al. 2002) and shear strength (Jiang et al. 2004, Thornton and Antony 2000, Richefeu et al. 2005).

In this paper, we investigate shear strength properties of wet granular materials as a function of water content in the pendular state. Sand and glass beads were wetted and tested in a direct shear cell. In parallel, we carried out DEM simulations by means of a 3D numerical code (tapio-K) based on molecular dynamics approach developed in our laboratory. We used an explicit expression of the capillary force as a function of interparticle distance, water bridge volume and surface tension. A peculiar feature of our system is that the effects of capillary bonding can be observed only if the confining pressure is of the same order of magnitude or below the tensile strength or the Coulomb cohesion. The latter is proportional to the strength of capillary bonds and inversely proportional to the particle size (Richefeu et al. 2006). For fine sand, typical values are below 1 kPa, imposing us to work with low confining pressure.

We find good agreement between experiments and numerical simulations. We show that the shear strength is mostly controlled by the distribution of liquid bonds. This property results in the saturation of shear strength as a function of water content. We arrive at the same conclusion by analyzing the shear strength from the microstructure and by accounting for grain polydispersity. Finally, we discuss the applicability of the discrete element approach to unsaturated soils under more general loading paths.

2 Discrete Element Method

We developed a DEM code based on the Molecular Dynamics (MD) method (Cundall and Strack 1979, Allen and Tildesley 1987). The grain motions are explicitly integrated by accounting for their interactions (e.g., contact, friction, capillary cohesion, magnetisation ...). Here, we present the interaction laws used for the contact, friction and capillary cohesion in the case of spherical grains.

2.1 Contact and Friction Laws

The normal contact law is modeled as a spring and a dashpot in parallel. The normal force f_n^{ctc} is given by

$$f_n^{ctc} = -k_n \delta_n + \gamma_n v_n \,, \tag{1}$$

where δ_n is the normal distance (negative or zero when the grains are in contact), k_n is the normal stiffness, v_n is the relative normal velocity, and γ_n is a damping coefficient.

The friction law expresses the tangential force as a function of the sliding velocity $\mathbf{v}_s$. The friction force $\mathbf{f}_t$, contained in the contact plane, obeys a viscous-regularized Coulomb law:

$$\mathbf{f}_t = \min\{\gamma_t \mathbf{v}_s,\ \mu f_n^{ctc}\} \frac{\mathbf{v}_s}{\|\mathbf{v}_s\|} , \tag{2}$$

where γ_t is the tangential viscosity and μ is the coefficient of friction.

2.2 Capillary Cohesion Law

In the presence of water, the grains can be connected by a liquid bridge as illustrated in Fig. 1**a**. The liquid bridge gives rise to an attractive force, the force of capillary cohesion, between the grains. This force results from the combined action of the difference in pressure between the liquid and gas phases and the liquid surface tension γ (Pierrat and Caram 1997, Willett et al. 2000, Gröger et al. 2003, Soulié et al. 2006). We use the explicit expression of the capillary law of cohesion proposed by Soulié et al. (2006) in the polydisperse case (Fig. 1**b**). This law, expressing the capillary force f_c as a function of the normal distance δ_n, is given by

$$f_c = \begin{cases} -\pi\gamma\sqrt{R_1 R_2}\{\exp(A\frac{\delta_n}{R_2} + B) + C\} & \text{for } \delta_n > 0 \\ -\pi\gamma\sqrt{R_1 R_2}\,\{\exp(B) + C\} & \text{for } \delta_n \le 0 \end{cases} \tag{3}$$

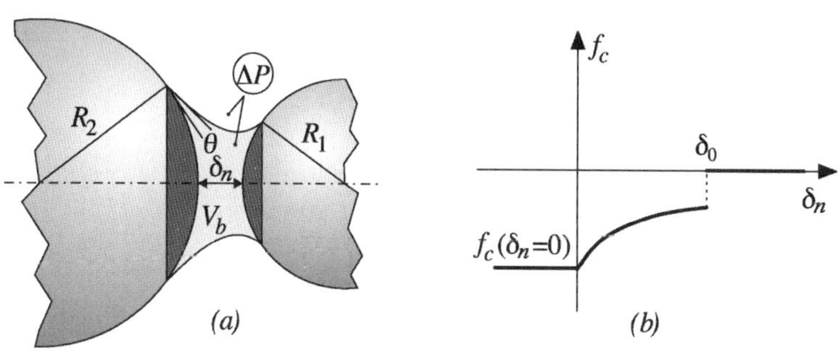

Fig. 1. Capillary cohesion law. (**a**) Physical and geometric parameters of a capillary bond. (**b**) Typical behavior of the capillary strength as a function of the normal distance

where R_1 and R_2 are the grains radii ($R_1 < R_2$); A, B and C are parameters which depend on the volume V_b of the liquid bond and on the contact angle θ; δ_0 is the rupture distance. From a theoretical study of a liquid bridge between two spheres, Lian et al. (1993) proposed the following rupture criterion:

$$\delta_0 = (1 + 0.5\theta)V_b^{1/3} . \tag{4}$$

We used this criterion which was validated by local experiments (Soulié 2005). It is also necessary to define a criterion for the formation of a liquid bond. It is considered that a liquid bridge can be formed (or to reform itself) when the grains are in contact ($\delta_n \leq 0$). At the numerical level, the main difficulty comes from the management of the distribution of the total volume of water between the grains, i.e. the determination of local volumes of the capillary bonds. In the simulations presented in this paper, the total volume of water is distributed over all the possible bonds of the sample in proportion to the radii of the touching grains. This corresponds to a global and homogeneous management of the distribution of water, which makes it possible to connect the local water volumes to the macroscopic water content. Local water redistributions were considered, too, see Richefeu (2005). The distribution protocol plays a determining role in the mechanical resistance of material, as we shall see below. However, this information remains difficult to reach in experiments with real materials.

3 Direct Shearing at Low Confining Stresses

Several modifications were made to the standard Casagrande's shear box in order to be able to measure the shear strength of granular materials to very low confining pressures (lower than 1 kPa). We present here the main experimental and numerical results obtained. These results will be analyzed through a theoretical expression of the Coulomb cohesion resulting from a microscopic analysis.

3.1 Experiments and Simulations

Figure 2 shows a schematic view of the setup, its dimensions, and certain notations used. A normal pressure $\sigma = g(m_N/S + \rho\ h)$ is applied to the sample by means of a flask whose mass m_N can be modified by pouring sand into it. ρ is the bulk density and g is the gravity. A shear stress $\tau = m_T/S$ is incrementally imposed by gradually filling a cup connected to the moving part of the box by a system of pulley. The stress reached during the rupture of the sample is the yield shear stress. Several tests carried out with various normal stresses allow us to estimate the yield locus in the Mohr–Coulomb plane. In direct shearing, the boundary conditions are such that the finite size effects cannot be fully eliminated (Thornton and Zhang 2001). We did not

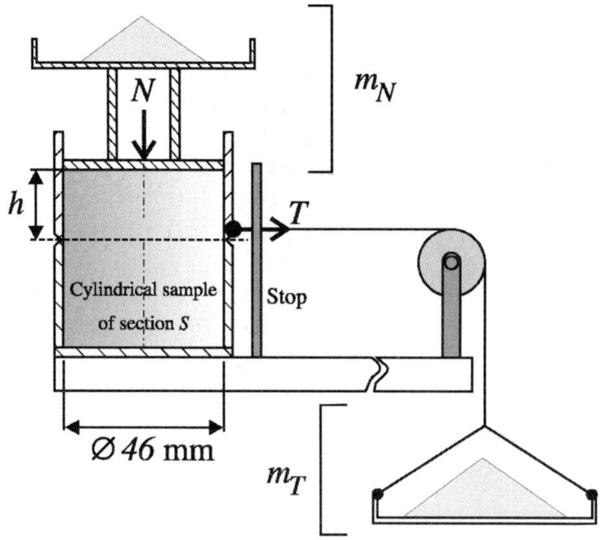

Fig. 2. Experimental Setup

check such size effects in our experiments. Given the system size, this requires using considerably more particles with smaller size, affecting thus the level of Coulomb cohesion in the presence of water. But we checked that the measured strengths are reproducible and in good agreement with analytical estimates (see below).

Figure 3**a** shows the yield loci obtained for a sand made up of angular grains of diameters ranging from 0.1 to 0.4 mm with various water contents w

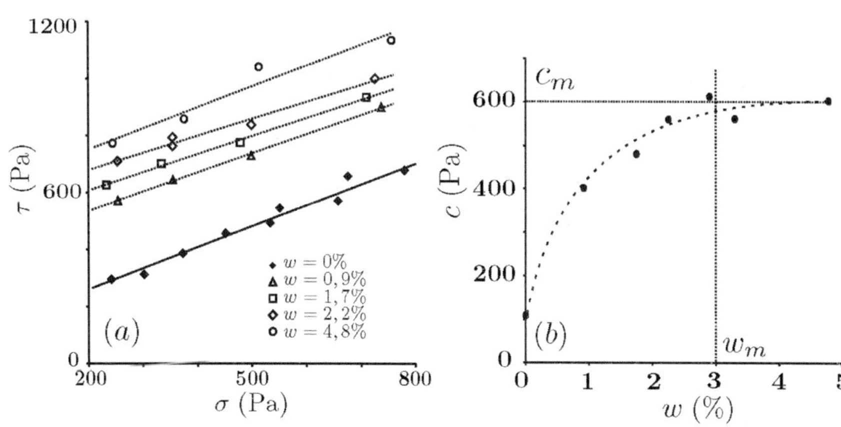

Fig. 3. Experimental results. (**a**) Yield loci estimated for a sand. (**b**) Coulomb cohesion as a function of water content

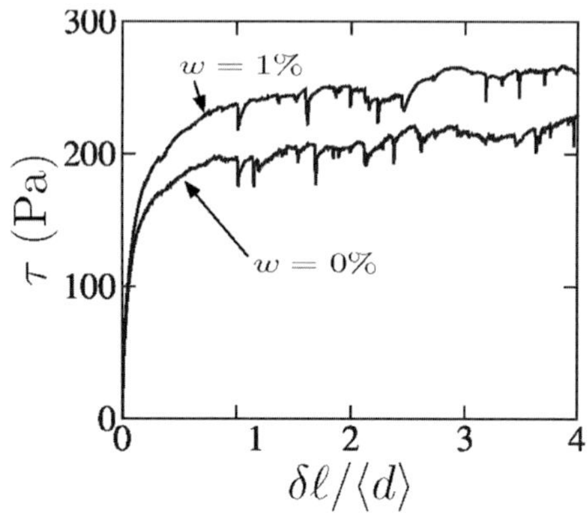

Fig. 4. Stress-strain relations for $w = 0\%$ and $w = 1\%$ ($\sigma = 300\,\mathrm{Pa}$)

(ratio of the water mass on grains mass). One can notice that the experimental points are well fit to straight lines, in agreement with the Mohr–Coulomb model $\tau = \tan\varphi\,\sigma + c$ characterized by two parameters: the macroscopic cohesion c and the internal friction angle φ. This figure also shows the good reproducibility of the experiments. The angle φ does not seem to vary appreciably with w in contrast to the parameter of cohesion which increases in a nonlinear way with water content and is saturated to a value c_m for a water content w_m (Fig. 3**b**). For the water contents above 5%, we leave the pendular regime.

The numerical tests are similar to the experimental tests (same level of loading, same dimensions of the box . . .). A numerical sample of 7307 spherical grains was prepared with a solid fraction $\phi = 0.6$. The number of particles is imposed by the system dimensions and particle size in order to reach the same level of packing fraction as in experiments. Various water contents were obtained by distributing the total water volume over all the contacts ($\delta_n \leq 0$) in this sample. A set of 15 simulations (three levels of the normal stress and five values of the water content) were performed. Figure 4 presents the stress-strain plots obtained for $w = 0\%$ and $w = 1\%$. The rupture stress is considered to be the residual shear stress because the shear stress approaches the residual state without passing by a peak state. As for the experiments, the evolution of the macroscopic cohesion can be analyzed as a function of the water content. Figure 5**a** shows the estimates of yield loci obtained numerically. One can note that the numerical model reproduces well the experimental data. The water content does not have an influence on the internal angle of friction and the cohesion parameters increases in a nonlinear way and saturates at a relatively low level of the water content (Fig. 5**b**).

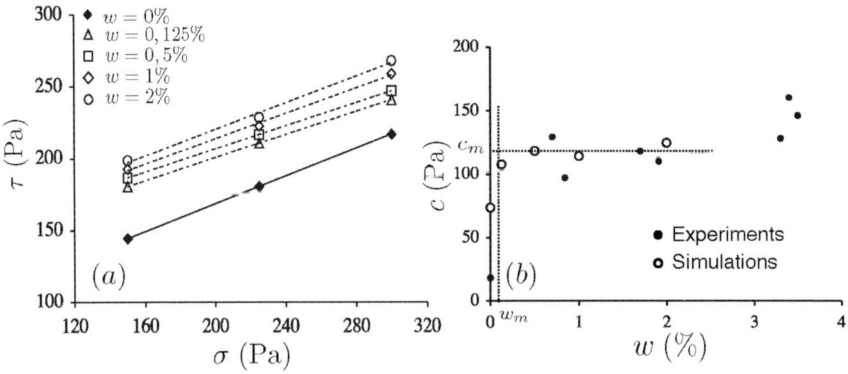

Fig. 5. Simulation results. (**a**) Estimated yield loci. (**b**) Coulomb cohesion as a function of water content (experiments were carried out with 1 mm glass beads)

3.2 Micromecanical Analysis

From a micromechanical analysis, a theoretical expression of the Coulomb cohesion was established (Richefeu et al. 2006):

$$c^{th} = \frac{3}{2\pi}\mu\kappa s \frac{\phi z}{\langle D \rangle}, \tag{5}$$

where κ is a parameter that depends essentially on the surface tension of water, s is a factor related to the polydispersity of material, z is the average number of liquid bonds per grain, and $\langle D \rangle$ is the average diameter of the grains. The validity of this expression was shown by comparing the predicted theoretical values of Coulomb cohesion with experimental and numerical measurements (Table 1). This comparison shows also that the numerical results are in quantitative agreement with the experiments.

In Table 1, we show also experimental results of three other materials (Richefeu et al. 2006, Richefeu 2005): (1) tightly-graded polydisperse glass

Table 1. Measured and theoretical parameters for the experimental and numerical samples

	Sand	GB1	GB2	GB3	Simulations
$\langle D \rangle$ (mm)	0.16	0.45	0.60	1.00	1.65
s	0.50	0.99	0.91	1.00	0.79
z	6	6	6	6	9
ϕ	0.6	0.6	0.6	0.6	0.6
$\tan\varphi$	0.66	0.58	0.58	0.46	0.48
c_m (Pa)	600	350	300	150	120
c_{th} (Pa)	709	438	302	158	118

beads with diameters from 0.4 mm to 0.5 mm (denoted by GB1), (2) well-graded polydisperse glass beads with diameters from 0.4 mm to 0.8 mm (denoted by GB2), and (3) monodisperse glass beads of diameter 1 mm (denoted by GB3).

4 Conclusions

Experiments and the simulations based on discrete element simulations were used with the aim of studying and analyzing the shear strength of unsaturated granular materials in the pendular state at low levels of loading. It was shown that the Coulomb cohesion increases with the water content and that it is saturated to a maximum value which depends only on the nature of the material. A theoretical analysis based on micromechanics allows us, on one hand, to validate the numerical modeling and, on the other hand, to show that the increase in the shear strength with water content is primarily due to an increase in the number of liquid bonds per unit volume.

In the extension of this work, it is essential to evaluate the limits of the model by considering other materials and non monotonous loading paths. In particular, we would like to study the shear strength of granular media with a larger polydispersity than materials that were used in the present investigation. Since the distribution of liquid bonds seems to be a major parameter for the cohesion of unsaturated granular materials, it also merits to be investigated in more detail experimentally. Finally, an interesting application of the ideas put forward in this paper would be to examine by which mechanisms the cohesion of a sample of unsaturated sand increases as a result of compactification.

References

Allen MP, Tildesley DJ (1987) Computer Simulation of Liquids. Oxford University Press, Oxford

Bardet JP, Proubet J (1991) Geotechnique 41(4):599–613

Bika D, Gentzler M, Michaels J (2001) Powder Technology 117:98–112

Cundall PA, Strack ODL (1979) Geotechnique 29:47–65

Gröger T, Tüzün U, Heyes DM (2003) Powder Technology 133:203–215

Iveson S, Beathe J, Page N (2002) Powder Technology 127:149–161

Jiang MJ, Leroueil S, Konrad JM (2004) Computers and Geotechnics 31:473–489

Lian G, Thornton C, Adams MJ (1993) Journal of Colloid and Interface Science 161:138–147

Pierrat P, Caram H (1997) Powder Technology 91:83–93

Richefeu V (2005) Approche par éléments discrets 3D du comportement de matériaux granulaires cohésifs faiblement contraints. PhD Thesis, Université Montpellier 2, France

Richefeu V, El Youssoufi MS, Peyroux R, Bohatier C (2005) Frictional contact and cohesion laws for Casagrande's shear test on granular materials by 3D DEM – comparison with experiments. In: García-Rojo R, Herrmann HJ, McNamara S (eds) Powders and Grains 2005. A.A. Balkema Publisher, pp. 509–512.
Richefeu V, El Youssoufi MS, Radjaï F (2006) Physical Review E 73:051304
Soulié F (2005) Cohésion par capillarité et comportement mécanique de milieux granulaires. PhD Thesis, Université Montpellier 2, France
Soulié F, Cherblanc F, El Youssoufi MS, Saix C (2006) Int J Numer Anal Meth Geomech 30:213–228
Staron L, Vilotte JP, Radjaï F (2002) Physical Review Letter 89(20):204302
Thornton C, Antony SJ (2000) Powder Technology 109:179–191
Thornton C, Zhang L (2001) A DEM comparison of different shear testing devices. In: Kishino (ed) Powders and Grains 2001. A.A. Balkema Publisher, pp. 183–190
Willett C, Adans MJ, Johnson S, Seville J (2000) Langmuir 16:9396–9405

Geochemical Effects on Swelling Pressure of Highly Compacted Bentonite: Experiments and Model Analysis

Mingliang Xie[1], Helge C. Moog[2], and Olaf Kolditz[1]

[1] GeoSystemsResearch, Center for Applied Geoscience, University of Tübingen, Sigwartstr. 10, D-72076 Tübingen, Germany
mingliang.xie@uni-tuebingen.de, kolditz@uni-tuebingen.de
[2] Gesellschaft für Anlagen- und Reaktorsicherheit mbH (GRS), Theodor Heuss Str. 4, D-38122 Braunschweig, Germany Helge.Moog@grs.de

Summary. Bentonite is widely selected to to be used as buffer material for high-level nuclear waste (HLW) repositories owing to its favorite hydrogeological and geochemical properties. This is because mainly of its moisture swelling effect. Experimental and theoretical evidences indicate that the swelling characteristic is largely influenced by the porewater chemistry. A chemical swelling model for constrained condition is developed on the basis of diffuse double layer (DDL) theory and related the microscopic theory to the macroscopic swelling pressure. Experiments with purified clay fraction ($< 2\,\mu m$) of MX–80 bentonite were undertaken. The fine bentonite was compacted to a dry density of $1600\,kg/m^3$ with initial liquid saturation of 35.7% and then installed into a rigid container for swelling pressure experiment. The unsaturated bentonite sample was then flushed with NaCl solutions in different concentrations. With the increase of the ionic strength, the measured swelling pressure decreases. The experimental swelling pressure values agree well with the modelled results using the chemical swelling model.

Key words: geochemistry, bentonite, swelling pressure, experiment, model analysis

1 Introduction

Compacted bentonite or bentonite/sand mixtures are widely selected to be used as buffer materials for waste disposal, especially for high level radioactive waste (HLW) repositories. This is due to their favorable properties, especially the remarkable swelling capability of bentonite, which restricts migration of water and contaminants due to self-sealing effect. Bentonites, which consist to a considerable extent of swelleable clay minerals, swell and shrink with wetting and drying processes. The magnitude of swelling and shrinking properties of bentonite can be greatly influenced by many factors, which can be divided into two groups – factors that depend on the nature of the soil particles (type and amount of clay minerals) and factors determined by the placement as well as environmental conditions (dry density, initial water content, compaction method, soil structure, thermal conditions, electrolyte concentration in the pore water) (Norrish 1955, Seed et al. 1962, Low 1979, 1987, Huyghe and Janssen 1999). The swelling effect results from additional embedding of water/solution molecules into the soil matrix, especially into the interlayers of expansive minerals. In the case of free swelling, volume changes up to 1400 to 2000% for purified dry Na-montmorillonite samples in powder immersed in pure water and changes up to 45 to 145% for Ca-montmorillonite are reported (Mielenz and King 1955). In the case of constrained swelling, however, swelling can result in macro- and mesopore volume reduction and simultaneously in swelling pressure, which can be measured in oedometers or triaxial cells.

The origin of bentonite swelling has been widely studied by many researchers, in efforts to derive a fundamental approach relating swelling potential with basic particle-water-cation interactions (see e.g. Miller and Low 1990, Low 1992, Quirk 1997, Iatridis et al. 2003). These approaches are based on net negative total charge on the surface of montmorillonite particles, the Poison–Boltzmann (PB) theory, the Gouy–Chapman Diffuse Double Layer (DDL) theory (see e.g. Gouy 1910, Chapman 1913) and the DLVO theory (named after Derjaguin, Landau, Verwey and Overbeek) (see e.g. Derjaguin and Landau 1941, Verwey and Overbeek 1948). However, they are restricted to the microscopic level and the use of semi-empirical equations derived from experimental data under free swelling conditions. Therefore, limit the applicability of the models to describe the complex coupled processes. A connection between the microscopic theory to the macroscopic effects is still important for modelling purpose.

Classical modelling of the swelling behavior of bentonite was developed on the basis of laboratory oedometer tests and the concept of Terzaghi's effective stress in order to obtain the stress-strain behavior of expansive clays (see e.g. Mitchell 1993, Sridharan 1990). Numerical modelings of complex THMC processes in bentonite as buffer materials have been developed for design of waste repositories based on essential aspect of coupled processes occurring within the material (see e.g. Börgesson 1985, Bennethum and Cushman 2002). The first model to evaluate those coupled processes was suggested by Noorishad et al. (1982).

This paper provided a new chemical swelling model for describing the constrained swelling pressure with pure bentonite. Experiments using purified MX80 bentonite were undertaken by flushing with different concentrations NaCl solutions. Model calculations of experimental results agree well.

2 Chemical Swelling Model

2.1 Conceptual Model

Bentonite is naturally occurring material which is predominantly composed of smectitic minerals (montmorillonite, beidellite, nontronite) (Jasmund and Lagaly 1993), in which smectite plays a central role for swelling. Typical smectite like montmorillonite is composed of two structural units, the silica tetrahedron and the alumina octahedron. These units are interconnected and extend to form thin sheets (each is only about 14 Å in thickness). Several such sheets are stacked one above the other, to form a crystal particle. Owing to the weak connection between some of the adjacent sheets (active layers), water easily enters into them with the help of electron repulsion, which results from the net charge of the bentonite particle. Consequently, the particles may be separated into smaller pieces. Katti and Shanmugasundaram (2001) observed the breakdown of the clay agglomerates into small sized particles by increasing saturation through digital analysis of the images obtained by using energy-dispersive X-ray analysis. Herbert and Moog (2002) proved variation of the interlayer distances by changing porewater chemical compositions.

2.2 Interlayer Porosity (Microscopic)

According to the free swelling model based on the diffuse double layer (DDL), the total porosity n of bentonite can be distinguished into interlayer n_{IL} (porosity within the particles) and interparticle porosity n_{IP} (porosity between particles) (Xie et al. 2004)

$$n = n_{\text{IP}} + n_{\text{IL}} \,. \tag{1}$$

With this definition, n_{IP} reveals the coarse pore and can be assumed as the initial porosity of a dry sample. In the case of constrained swelling, the total porosity remains constant. The interlayer porosity increases with the saturation process. Therefore, interparticle porosity decreases accordingly.

Around the particles and within the interlayer space, diffuse double layers tend to build up, completely only in the case of free swelling. The maximum interlayer porosity for one particle with m effective layers can be calculated as follows (Xie et al. 2004):

$$n_{\text{IL}\max} = mS_0\rho_d \left(\frac{\varepsilon\varepsilon_0 RT}{2F^2 I}\right)^{\frac{1}{2}} \,. \tag{2}$$

Taking account of the fact that swelling pressure increases with the saturation process and that the magnitude of swelling potential depends on the expansive fraction, the total interlayer porosity n_{IL} of the specimen can be calculated:

$$n_{\text{IL}} = (S^l)^\eta \cdot \beta \cdot n_{\text{IL}\max} \tag{3}$$

in which S^l is degree of liquid saturation, β represents volume fraction of the expansive minerals, η is a dimensionless coefficient.

2.3 Swelling Pressure Model

Swelling pressure of bentonite is fundamentally a sort of internal energy of the expansible minerals Xie et al. (2007), it is generally measured under constrained condition with the intrusion of solutions. The concept of swelling potential was firstly used by Seed et al. (1962) to represent the swelling ability of bentonite. In the present paper, swelling potential is defined as the porosity change potential n_{sw} responsible for swelling pressure equation (4). In the case of constrained swelling, interlayer porosity change (Δn_{IL}, which is the difference between the current and initial interlayer porosity) can not fully be converted into swelling pressure. Part of it is used to compact the large pores (initial interparticle porosity n_{IP0}), especially at the early stage. The looser the sample, the more interlayer porosity change will be converted. Theoretically, the large pores can totally be compacted, but in reality, a certain interparticle porosity always remains. This is defined as the minimum porosity ($n_{IP\min}$). This value is lower, the higher the swelling pressure σ. It depends on the compressibility of the material.

$$n_{sw} = \Delta n_{IL} - (n_{IP0} - n_{IP\min}), \tag{4}$$

$$n_{IP0} = n_0 - n_{IL0}, \tag{5}$$

$$n_{IP\min} = f(\sigma, I) \tag{6}$$

where n_0 is the initial porosity, n_{IL0} is the initial interlayer porosity.

To calculate the swelling pressure measured in laboratory experiments, the following assumptions are made:

1. The sample is unsaturated initially, and fully saturated at the end;
2. The sample is homogenous and isotropic;
3. Geochemical reactions (e.g. ionic exchange) is at equilibrium state;
4. The effective expansive layer number of each particle is dependent on ionic strength of the pore water solution, $m = f_1(I)$;
5. The $n_{IP\min}$ value is dependent on swelling pressure for one experiment. To consider the chemical effect on the swelling pressure, this value is assumed to be dependent on ionic strength of the pore water solution, $n_{IP\min} = f_2(I)$. The higher the concentration, the lower the swelling pressure, thus the higher the $n_{IP\min}$ value.

The swelling strain ϵ_{sw} is a volumetric strain and equal to the swelling potential. The swelling pressure P_{sw} is a volumetric stress, which can be calculated from the swelling strain:

$$P_{\min} = K'\epsilon_{sw} = K'n_{sw} \tag{7}$$

in which K' is the volumetric modulus, which can be calculated (IGS (1998)):

$$K' = \frac{E}{3 - 6\nu} \tag{8}$$

where E is the elastic modulus and ν is the poisson ratio.

3 Experiment

In order to investigate the chemical effects on the swelling pressure, a series of experiments were undertaken with MX–80 bentonite under the intrusion of NaCl solutions in various concentrations. The bentonite was purified and treated by dilute NaCl solution. The fine fraction (< 2 µm) of MX–80 bentonite which mainly consists of montmorillonite was separated from the bulk material by sedimentation and converted into the Na-form. After freeze drying the material was compacted to a dry density of 1600 kg/m^3 in cylinder form (5 cm in diameter and 2.4 cm in hight). Initial liquid saturation is 35.7%. The concentrations of NaCl solutions are 0.011 M, 0.0256 M, 0.116 M, 0.575 M and 1.053 M, respectively.

To measure the maximum swelling pressure, the sample was installed in a rigid cylindrical container and no volume change is thus allowed. Constant Injection pressure of 25 MPa was applied. Density of outflowing solution was checked regularly. When density approached the expected value for the inflowing solution, the percolate was analyzed. Inlet- and outlet-valves were opened and the internal pressure recorded. Total pressure at equilibrium was taken as swelling pressure. Afterwards the sample was connected to a new solution with different ionic strength and the described procedure repeated. Total equilibration time at each step amounted to 4–5 months.

The measured results are shown in Fig. 1. It is clear to see that the swelling pressure decreases with the increase of ionic strength of the NaCl solution.

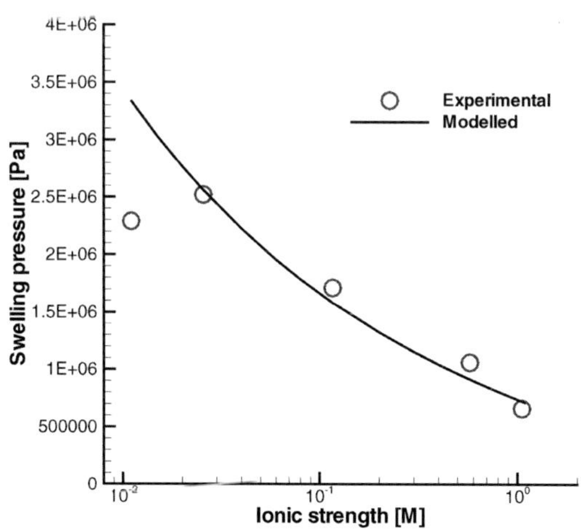

Fig. 1. Comparison of the experimental and modelled swelling pressure of bentonite and its dependence on ionic strength

4 Model Calculation

The measured swelling pressure results were calculated using the chemical swelling model described in section 2. Some parameters used are shown in Table 1. The effective layer number m in equation (2) is:

$$m = m_0 I^a \tag{9}$$

in which m_0 is the maximum expansive layer number, a is a parameter fitted for the experiment to be 0.19.

Table 1. Porous medium properties

Symbol	Meaning	Value/Ref	Unit
ρ_d	density	1600	kg/m^3
n_0	porosity	0.37	–
S_0	extern. spec. surface	31.80	m^2/g
β	swelling fraction	1.00	–
η	coefficient in eqn (2)	1.50**	–
m_0	maximum effective layer number	22*	–
S_0^l	initial saturation	0.357	–
S_{sat}^l	fully saturated saturation	1.0	–
n_{min0}	min. porosity	0.03	–
E	Young's modulus	3.5	MPa
ν	Poisson ratio	0.3	–

* Estimated value according to FEBEX Working groups (2000)
** From Xie et al. (2004)

The minimum porosity n_{min} is also assumed to be dependent on ionic strength as described in section 2.

$$n_{min} = n_{min0} I^b \tag{10}$$

in which b is a parameter fitted to be 0.5.

The modelled results are compared with the experimental value as shown in Fig. 1. The modelled results agree well with the experimental data. If the ionic strength is over 1.0, the swelling pressure decreases slowly, e.g. the changing rate decreases also.

5 Conclusions and Remarks

This paper presents the experimental and theoretical calculation to investigate the geochemical effect (ionic strength) on the swelling pressure of highly compacted pure bentonite under constrained conditions. The measured data showed that with the increase of ionic strength of the injected NaCl solutions, the measured swelling pressure decreases, which coincides well with the theoretical and experimental data in literature (e.g. Norrish 1955, Herbert and Moog 2002).

This results were recalculated with a chemical swelling model. The effective layer number and minimum porosity are dependent on the ionic strength. Further experiments including ionic exchange and chemical reactions (e.g. dissollution/precipitation) are still undertaken.

Acknowledgement

The work is supported by German Federal Ministry of Education and Research (BMBF) under grant 02C1295. The authors would like also to acknowledge Hua Shao (BGR Hannover), Tom Schanz (University of Weimar), Horst-Jürgen Herbert (GRS Braunschweig) and Jens Birkholzer (Lawrence Berkeley National Laboratory) for providing suggestions and very useful discussions to this research. We thank Dr. Pitterich (FZK) for his support of this research work.

References

Bennethum L, Cushman J (2002) Multicomponent, multiphase thermodynamics of swelling porous media with electroquasistatics II: Constitutive theory, Transport Porous Med 47:337–362

Bennethum L, Weinstein T (2004) Three pressures in porous media, Transport Porous Med 54:1–34

Börgesson L (1985) Water flow and swelling pressure in non-saturated bentonite-based clay barriers, Eng Geol 314:229–237

Chapman D (1913) A contribution to the theory of electrocapillarity, Philosophical Magazine 25:475

Derjaguin B, Landau L (1941) Theory of the stability of strongly charged lyophobic soils and of the adhension of stronghly charged particles in solutions of electrolytes, Acta Physiochem USSR 14:633–662

FEBEX Working groups (2000) Full-scale engineered barriers experiment for a deep geological repository for high level radioactive waste in crystalline host rock. Final report

Gouy G (1910) Sur la constitution de la charge électrique à la surface d´un électrolyte, Journal de Physique 9:457–468

Herbert HJ, Moog HC (2002) Untersuchungen zur Quellung von Bentoniten in hochsalinaren Lösungen. Abschlussbericht, Gesellschaft für Anlagen und Reaktorsicherheit (GRS) mbH

Huyghe J, Janssen J (1999) Thermo-chemo-electro-mechanical formulation of saturated charged porous solids, Transport Porous Med 34:129–141

Iatridis J, Laible J, Krag M (2003) Influence of fixed charge density magnitude and distribution on the intervertebral disc: Applications of a poroelastic and chemical electric PEACE model, J Biomech Eng 125:12–24

IGS (1998) Recommended descriptions of geosynthetics functions, geosynthetics terminology, mathematical and graphical symbols. Report, International Geosynthetics Society (1998)

Jasmund K, Lagaly G (eds) (1993) Tonminerale und Tone. Strukturen, Eigenschaften, Anwendungen und Einsatz in Industrie und Umwelt. Steinkopff Verlag, Darmstadt, ISBN-3-7985-0923-9

Katti D, Shanmugasundaram V (2001) Influence of swelling on the microstructure of expansive clays, Can Geotech J 38(1):175–182

Low PF (1979) Nature and properties of water in montomorillonite-water systems, Soil Sci Soc Am J 43:651–658

Low PF (1987) Structural component of the swelling pressure of clays, Langmuir 3:18–25

Low PF (1992) Interparticle forces in clay suspensions: Flocculation, viscous flow and swelling. In: Proc 1989 Clay Min Sco Workshop on Rheology of Clay/Water Systems

Mielenz RC, King ME (1955) Physical mechanical properties and engineering performance of clays. California Division of Mines Bulletin, Clays and clay technology: 1st National Conference on Clays and Clay Technology, Berkeley, California 169:196–254

Miller SE, Low PF (1990) Characterization of the electrical double layer of montmorillonite, Langmuir 6:572–578

Mitchell J (1993) Fundamentals of soil behaviour, 2nd edn. John Wiley & Sons

Noorishad J, Ayaollahi MS, Witherspoon P (1982) A finite-element method for coupled stress and fluid flow analysis in fractured rock masses., Int J Rock Mech Min Sci 19:185–193

Norrish K (1955) Manner of swelling of montmorillonite, Nature 4397:256–257

Quirk J (1997) Application of double-layer theories to the extensive crystalline swelling of Li-Montmorillonite, Langmuir 13:6241–6248

Seed HB, Woodward RJ, Lundgren R (1962) Prediction of swelling potential for compacted clays, J Soil Mechanics and Foundation Division 88:53–87

Sridharan A (1990) Strength and volume change behaviour of a sand-bentonite mixture, Can Geotech J 27:404

Verwey E, Overbeek J (1948) Theory of the stability of lyophobic colloids. Elsevier, Amsterdam

Xie M, Agus S, Schanz T, Kolditz O (2004) An upscaling method and numerical modelling of swelling/shrinking processes in compacted bentonite/sand mixtures, Int J Numer Anal Meth Geomech 28:1479–1502

Xie M, Wang W, Kolditz O (2007) Numerical modelling of swelling pressure in unsaturated expansive elasto-plastic porous media, Transport Porous Med (in print)

Use of State Surface Approach to Explain the Barcelona Basic Model

Xiong Zhang[1] and Robert L. Lytton[2]

[1] Department of Civil and Environmental Engineering, University of Alaska Fairbanks, AK99775, USA `ffxz3@uaf.edu`
[2] F.J. Benson Chair, Department of Civil Engineering, Texas A&M University, College Station, Texas 77843-3136, USA `r-lytton@civilmail.tamu.edu`

Summary. State surface approach is becoming much less popular nowadays since many elasto-plastic modellers consider though it is useful as a simplified method in some practical problems, it can not be used to explain complex stress path dependency for unsaturated soils. Consequently, nearly all researchers used "normal compression lines at different suction levels" to develop their elasto-plastic models. In this paper, state surface approach is used to duplicate the examples given in the BBM model. All the examples under isotropic conditions can be reproduced without any programming. It is found that state surface approach, when used correctly, can bring greater convenience in explaining unsaturated soil behavior and generate better understanding of unsaturated soil behavior including stress path dependency.

Key words: unsaturated soil, matric suction, net mechanical stress, State Surface, Barcelona Basic Model, elasto-plastic, yeild curve

Introduction

Matyas and Radhakrishna (1968) put forward the concept of constitutive (state) surfaces to relate void ratio and degree of saturation with the net normal stress p and the matric suction s. Fredlund and Morgenstern (1977) performed null tests and concluded that only two stress state variables are needed for the description of volume change for unsaturated soils. Based on the two stress state variable concept and the concept of constitutive surfaces, Fredlund and his coworkers developed a complete framework for volume change theory for unsaturated soils. Most of the work is based on the assumption that the soil is elastic. However, unsaturated soils often demonstrate irrecoverable behavior. Alonso et al. (1990) proposed the first elasto-plastic model for unsaturated soils, which is called the Barcelona Basic Model (BBM). It was considered that constitutive surface can not be used to handle the stress path dependency for unsaturated soils and therefore the use of constitutive surface was abandoned. Instead, "normal compression lines at different matric suction levels" were used to develop their model. This methodology was extensively accepted by many other elasto-plastic modellers (Cui and Delage 1996, Wheeler and Sivakumar 1995). At present the use of constitutive surfaces was considered as an empirical method. In this paper, the BBM is used to illustrate the ability of constitutive surface to deal with stress path dependency.

Constitutive Surface Used in the BBM

The "normal compression lines at different matric suction levels" used in the BBM are essentially two dimensional images of a special shape of void ratio state surface, which is also called the void ratio constitutive surface by Fredlund and Rahardjo (1993). The surface consists of three surfaces as shown in Fig. 1, an elastic surface AFIH, a plastic collapsible soil surface FIJG, and a plastic expansive soil surface HCJI. The mathematical expression of the elastic void ratio constitutive surface AFIH is as follows:

$$e = C_1 - \kappa \ln p - \kappa_s \ln(s + p_{at}) \tag{1}$$

Eq. (1) can be obtained by combining Eqs. 2 and 4 in the original BBM (Alonso et al. 1990). The plastic collapsible soil surface FIJG has the following mathematical expression:

$$e = C_2 - \kappa_s \ln\left(\frac{s + p_{at}}{p_{at}}\right) - \lambda(s) \ln\left(\frac{p}{p^c}\right) \tag{2}$$

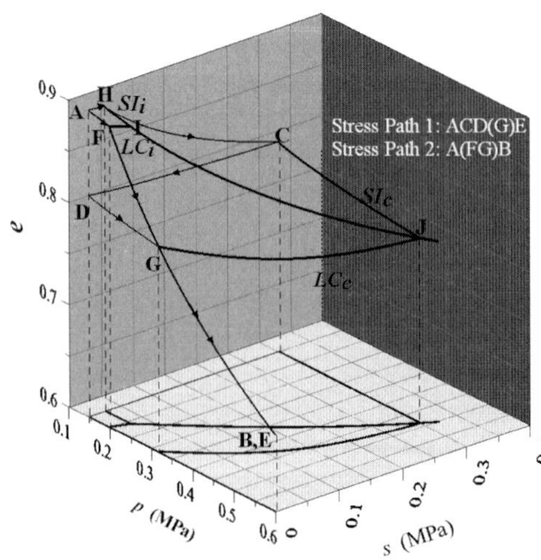

Fig. 1. Three dimensional plot of Fig. 14 in Alonso et al. (1990)

where $\lambda(s) = \lambda(0)[(1-r)\exp(-\beta s)+r]$. Eq. (2) can be obtained by combining Eqs. 1 and 6 in the original BBM. The plastic expansive soil surface HCJI has the following mathematical expression:

$$e = C_3 - \kappa \ln p - \lambda_s \ln(s + p_{at}) \tag{3}$$

where C_1, C_2, and C_3 are constants. Eq. (3) can be obtained by combining Eqs. 13 and 14 in the original BBM (Alonso et al. 1990). All the other symbols have the same definitions as those used in Alonso et al. (1990). The plastic collapsible soil surface and the plastic expansive soil surface are responsible for explaining the collapsible soil and expansive soil behavior, respectively. They form the elasto-plastic zone of the state surface.

Derivations of the LC and SI Curves

The LC yield curve is the intersection of the elastic surface and collapsible soil surface. In other words, the points on the LC curve must satisfy Eqs. (1) and (2) simultaneously, which gives

$$\ln\left(\frac{p}{p^c}\right) = \frac{C_4}{\lambda(s) - \kappa} \tag{4}$$

where C_4 is a constant. Eq. (4) is the projection of the LC curve on the $p : s$ plane. The LC curve has to pass through the yield point when the soil is saturated, that is, $p = p_0^*$ when $s = 0$, which gives:

$$C_4 = [\lambda(0) - \kappa] \ln\left(\frac{p_0^*}{p^c}\right). \tag{5}$$

Substituting Eq. (5) into Eq. (4), the LC curve for the BBM is obtained as follows:

$$p = p^c \left[\frac{p_0^*}{p^c}\right]^{\frac{\lambda(0)-\kappa}{\lambda(s)-\kappa}}. \tag{6}$$

Eq. (6) is exactly the same as the expression of the LC curve in the BBM.

In the same way, the SI yield curve is the intersection of the elastic zone and the expansive soil zone and the points on the SI curve must satisfy Eqs. (1) and (3) simultaneously, which gives

$$s = e^{\frac{C_3 - C_1}{(\lambda_s - \kappa_s)}} - p_{at} = \text{constant}. \tag{7}$$

Eq. (6) is the same as the expression of the SI curve in the BBM. Eq. (7) is the projection of the SI curve on the $p : s$ plane.

Evolution of the LC and SI Curves

The void ratio state surface has the same role as the consolidation curve for saturated soils. Therefore, the following rules should be followed:

1. The plastic (collapsible and expansive) surfaces always remain stationary (Eqs. (2) and (3) are unique). However, their borders with the elastic surface, which are the LC and SI curves, may change due to downward translation of the elastic surface.
2. When the stress state of the soil is in the elastic zone or soil experience an unloading from the LC or SI curve, all the soil behavior is recoverable and the elastic surface will remain stationary.
3. When the stress state of the soil is on the LC or SI curve, a further loading towards the plastic zone will cause the elastic surface to translate downwards (C_1 in Eq. (1) will decrease). New LC and SI curves are formed along the borders between the new elastic surface and the existing collapsible and expansive surfaces, respectively.

Reproducing the Examples Given in Alonso et al. (1990) by Using State Surface Approach

The examples used in Alonso et al. (1990) will be used to illustrate how the proper use of the "state surface approach" can bring convenience to the prediction of unsaturated soil behavior under isotropic loading conditions. The following soil parameters were used in the original BBM:

$$\begin{aligned} &\lambda(0) = 0.2; \quad \kappa = 0.02; \quad \lambda_s = 0.08; \quad \kappa_s = 0.008; \\ &r = 0.75; \quad \beta = 12.5\ \text{MPa}^{-1}; \quad p^c = 0.10\ \text{MPa}. \end{aligned} \tag{8}$$

Figures 2 and 3 show the three dimensional plots of Figs 11 and 12 (Case 1) in Alonso et al. (1990). For example, Figs 11**a** and 11**b** in Alonso et al. (1990) are the projections of Fig. 2 in the $p{:}s$ and $e{:}p$ planes, respectively. Both cases have the same initial conditions (point A), i.e. $p = 0.15\,\text{MPa}$, $s = 0.2\,\text{MPa}$, and $e_0 = 0.9$. By substituting parameters in Eqs. (8) and the initial conditions into Eq. (1) and considering the physical meanings of the LC and SI yield curves, it is found that C_1, C_2, and C_3 in Eqs. (1), (2), and (3) are 0.852, 1.042, and 0.786, respectively. The plastic expansive soil surface is not plotted in Figs 2, 3, or 4 since there is no stress path in that zone. All the stress paths are monotonic and isotropic loadings and will always fall on the state surface. The void ratio for any point on the stress paths can be directly calculated by substituting the corresponding stress state of the soil into Eq. (1) if the soil is in the elastic zone or Eq. (2) if the soil is in the plastic zone and the results are stress path independent.

Figure 4 shows the three dimensional plot of Fig. 13 in Alonso et al. (1990). In the second example of case 2 (Fig. 13), initial conditions $p = 0.15\,\text{MPa}$,

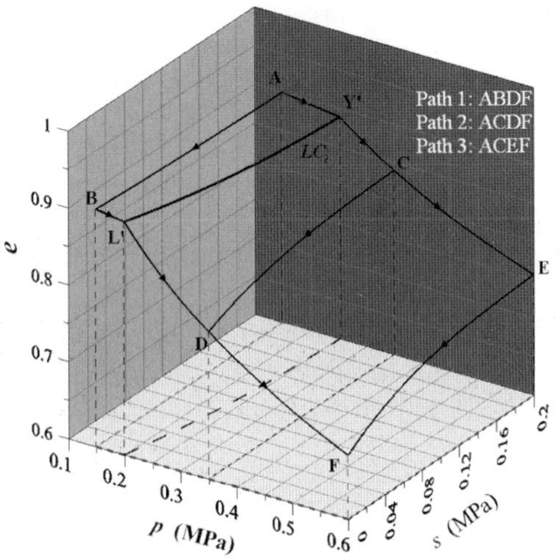

Fig. 2. Three dimensional plot of Fig. 11 in Alonso et al. (1990)

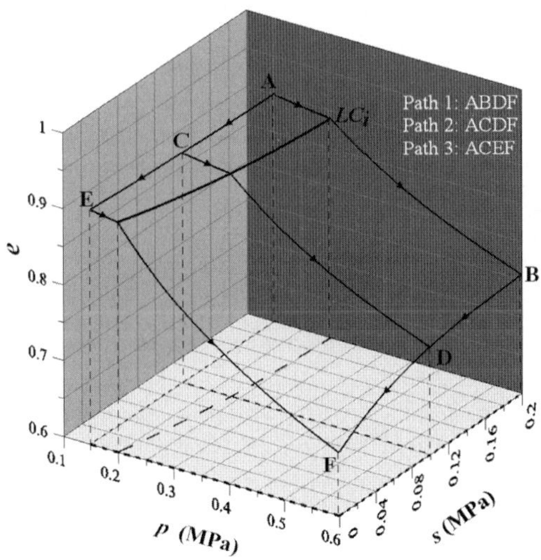

Fig. 3. Three dimensional plot of Fig. 12 in Alonso et al. (1990)

$s = 0.0\,\text{MPa}$, and $e_0 = 0.9$, $p_0^* = 0.2\,\text{MPa}$, and $s_0 = 0.3\,\text{MPa}$ give $C_1 = 0.844$, $C_2 = 1.033$ and $C_3 = 0.778$, respectively. The stress path ACEF is a monotonic and isotropic loading. Therefore, the corresponding specific volumes can be directly obtained from the constitutive surfaces by substituting the stress path to the corresponding surfaces. Stress path ABDF includes a monotonic and isotropic loading path AB and anunloading process BF. When the suction

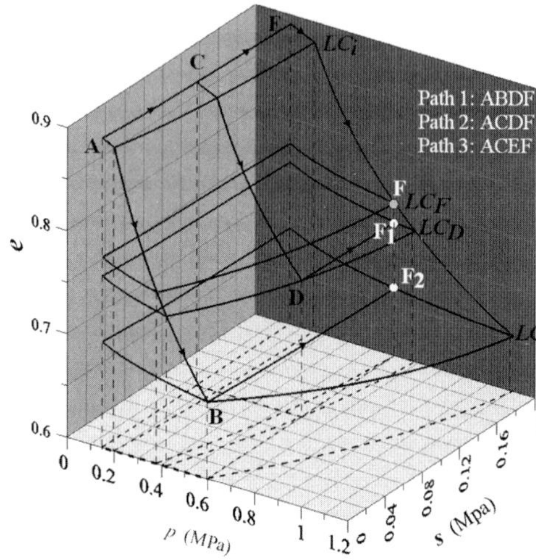

Fig. 4. Three dimensional plot of Fig. 13 in Alonso et al. (1990)

decreases from points B to F, a new unloading surface is generated along the LC_B yield curve and the corresponding C_1 in Eq. (1) is 0.703. Similarly, stress path ACDF includes a monotonic and isotropic loading path ACD and an unloading process DF. When the suction decreases from points D to F, a new unloading surface is generated along LC_D curve and the corresponding C_1 in Eq. (1) is 0.710. The void ratio for any point on the stress paths can be directly calculated by using the state surfaces and all the results are the same as that presented in the Alonso et al. (1990).

Fig. 1**a** shows the three dimensional plots of Figs. 14 in Alonso et al. (1990). The initial conditions $p = 0.15$ MPa, $s = 0.0$ MPa, and $e_0 = 0.9$, $p_0* = 0.20$ Mpa, and $s_0 = 0.025$ MPa give $C_1 = 0.844$, $C_2 = 1.033$, and $C_3 = 0.694$, respectively. The stress path ACDF includes a monotonically isotropic loading path AC, an unloading process CD and a monotonic reloading process DGF. As shown in Fig. 1**a**, the original LC and SI curves are FI and HI, respectively. Both the LC and SI curves expand due to the plastic loading stress path from points H to I. A new unloading surface CDGJ is generated when there is an unloading from point C to D and the corresponding C_1 in Eq. (1) is 0.760. As can be seen, C_1 in Eq. (1) decreases from 0.844 to 0.760. The preconsolidation stress of the soil increases due to the increase in the matric suction.

Conclusion

The constitutive surface has the same role as the normal consolidation curve for saturated soils. The BBM is based on a particular shape of state surface. The state surface approach can be used to explain the stress path dependency for unsaturated soils in an easier way. The use of state surface actually makes the derivation of the LC and SI yield curves much more concisely.

References

Alonso EE, Gens A, Josa A (1990) A constitutive model for partially saturated soils, Geotechnique 40(3):405–430

Cui YJ, Delage P (1996) Yielding and plastic behaviour of an unsaturated compacted silt, Geotechnique 46(2):291–311

Delage P., Graham I (1996) State of the Art Report-Understanding the Behavior of Unsaturated Soils Requires Reliable conceptual models. In: Proc 1st Int Conf on Unsaturated Soils, Paris 3:1323–1256

Fredlund DG, Morgenstern NR (1977) Stress state variables for unsaturated soils, ASCE J Geotech Eng Div GT5 103:447–466

Fredlund DG, Rahardjo H (1993) Soil Mechanics for Unsaturated Soils. John Wiley and Sons, New York

Matyas EL, Radhakrishna HS (1968) Volume change characteristics of partially saturated soils, Geotechnique 18(4):432–448

Wheeler SJ, Sivakumar V (1995) An elasto-plastic critical state framework for unsaturated soil, Geotechnique 45(1):35–53

Part II

Fundamental Methods

Saturated Elasto-Plastic Porous Media under Consideration of Gaseous and Liquid Phase Transitions

Wolfgang Ehlers and Tobias Graf

Institute of Applied Mechanics (CE), University of Stuttgart, 70 550 Stuttgart, Germany
`http://www.mechbau.uni-stuttgart.de/ls2`

Summary. Taking a closer look on, e.g., storage processes of greenhouse gases in deep geological aquifers or pressure changes in shear bands, the observation can be made that pressure and temperature changes in porous materials can induce phase transition processes of the respective pore fluids. For a numerical simulation of this behaviour, a continuum mechanical model based on a multiphasic formulation embedded in the well-founded framework of the *Theory of Porous Media* (TPM) is presented in this contribution. The single phases are an elasto-viscoplastic solid skeleton, a compressible pore gas consisting of the components air and gaseous pore water (water vapour) and an incompressible pore liquid, i.e., liquid pore water. The numerical treatment is based on the weak formulations of the governing equations, whereas the primary variables are the common temperature of the overall medium, the displacement of the solid skeleton and the effective pressures of the pore fluids. An initial boundary-value problem is discussed in detail, where the resulting system of strongly coupled differential-algebraic equations is solved by the FE tool `PANDAS`.

Key words: Theory of Porous Media (TPM), triphasic material, elasto-viscoplastic skeleton, viscous pore fluids, phase transitions between the pore fluids

1 Introduction

In geotechnical and environmental engineering, a coupled, non-isothermal approach of partially saturated soil is of growing interest, e.g., to estimate the effectiveness of the storage of greenhouse gases like carbon dioxide in geological formations. For a realistic modelling of such a process, phase transitions between the fluid phases have to be taken into account due to the fact that the gases are pressed at high temperatures into suitable formations. In this contribution, partially saturated soil is considered as a triphasic material within the *Theory of Porous Media* (TPM), where the model under study consists of a materially incompressible, elasto-viscoplastic solid skeleton saturated by two viscous fluid phases, a materially incompressible pore liquid (water) and a materially compressible pore gas composed of the components air and water vapour. For the isothermal case, the application of such triphasic model on the description of partially saturated soil has been discussed previously in Ehlers et al. (2004). Concerning the general TPM approach, the reader is referred, e.g., to the work by Ehlers (2002) and citations therein.

With regard to the organization of the following contribution, there are firstly the general kinematics and the underlying balance relations presented in Sect. 2. In Sect. 3, the required constitutive relations are identified. Therein, the soil matrix is governed by a general elasto-viscoplastic description of the solid stresses, whereas, as a result of the property of material incompressibility, the solid as well as the liquid material densities depend on the temperature. Concerning the pore fluids, the pressure of the materially incompressible liquid acts as a *Lagrange*an, whereas the pressures of the materially compressible fluid constituents are governed by the ideal gas law. Therefore, the resulting pore-gas phase is assumed as a mixture of ideal gases saturated with water vapour. The pore fluids interact by a capillary-pressure-saturation relation based on relative permeabilities to consider the mutually interacting pore fluid mobilities. The resulting system of differential-algebraic equations (DAE) is solved by use of the finite element tool PANDAS[1]. Finally, the injection process of heated air in a water saturated soil is considered as an initial boundary-value problem in Sect. 4 pointing out the capability of the triphasic formulation to simulate the coupled behaviour of unsaturated soil under non-isothermal conditions.

2 Basic TPM Equations Applied to a Triphasic Model

2.1 Kinematics and Balance Equations

In the frame of the TPM, a macroscopic model φ of superimposed and interacting continua φ^α can be formulated, where the assumption is made that all individual constituents are in a state of ideal disarrangement. In the further contribution, the index α indicates the following constituents: $\alpha = S$ (solid), $\alpha = L$ (liquid), $\alpha = V$ (water vapour) and $\alpha = A$ (air). In order to determine the volume fraction of the solid phase φ^S, the liquid phase φ^L and the gas phase $\varphi^G = \varphi^A + \varphi^V$, the volume fractions n^S and n^β, where $\beta = \{L, G\}$, are defined as the ratio of the volume elements dv^S and dv^β with respect to the volume fraction dv of the overall mixture φ at a local point. Assuming fully saturated conditions, the saturation constraint yields

$$n^S + n^F = 1 \qquad \text{with} \qquad n^F = n^L + n^G\,, \tag{1}$$

where n^F is the volume fraction of the overall fluid or the porosity, respectively. Additionally, the saturation functions $s^\beta := n^\beta / n^F$ can be introduced.

[1] **P**orous media **A**daptive **N**onlinear finite element solver based on **D**ifferential **A**lgebraic **S**ystems (http://www.get-pandas.com).

By the volume fractions n^α, two different density functions of each constituent φ^α can be defined and related to each other via

$$\rho^\alpha = n^\alpha \rho^{\alpha R} \quad \text{and} \quad \rho^\gamma = n^G \rho^\gamma_G , \quad \text{where} \quad \gamma = \{A, V\} . \tag{2}$$

The material (realistic or effective) density $\rho^{\alpha R}$ relates the local mass dm^α to the volume element dv^α, whereas the partial (global or bulk) density ρ^α relates the same mass to the volume element dv. Note that due to the fact that the overall gas phase is assumed as a mixture of ideal gases, the local mass of the air and water vapour can also be related to the volume of the gas phase: $\rho^\gamma_G = dm^\gamma / dv^G$. Furthermore, the solid and liquid phases are considered as materially incompressible, i.e., their material densities are only functions of the temperature θ.

Each spatial point $\mathbf{x}$ of the current configuration is at any time t simultaneously occupied by material particles (material points) P^α of all constituents φ^α. These particles proceed from different reference positions $\mathbf{X}_\alpha$ at time t_0. Thus, each constituent is assigned its own motion function

$$\mathbf{x} = \boldsymbol{\chi}_\alpha(\mathbf{X}_\alpha, t) . \tag{3}$$

It follows from (3) that each constituent has its own velocity and acceleration fields. The material time derivative, e.g., given for a scalar function Γ following the motion of φ^α, reads $(\Gamma)'_\alpha = \partial\Gamma/\partial t + \operatorname{grad}\Gamma \cdot \overset{\prime}{\mathbf{x}}_\alpha$. Therein, $\overset{\prime}{\mathbf{x}}_\alpha$ is the velocity of φ^α and the operator "grad $(\,\cdot\,)$" denotes the partial derivative of $(\,\cdot\,)$ with respect to the local position $\mathbf{x}$.

In the framework of the TPM, the partial balance equations of mass, linear momentum and energy hold for each constituent:

$$\begin{aligned}
(\rho^\alpha)'_\alpha + \rho^\alpha \operatorname{div} \overset{\prime}{\mathbf{x}}_\alpha &= \hat{\rho}^\alpha , \\
\rho^\alpha \overset{\prime\prime}{\mathbf{x}}_\alpha &= \operatorname{div} \mathbf{T}^\alpha + \rho^\alpha \mathbf{g} + \hat{\mathbf{p}}^\alpha , \\
\rho^\alpha (\psi^\alpha - \theta \eta^\alpha)'_\alpha &= \mathbf{T}^\alpha \cdot \mathbf{L}_\alpha - \operatorname{div} \mathbf{q}^\alpha + \rho^\alpha r^\alpha + \hat{\varepsilon}^\alpha .
\end{aligned} \tag{4}$$

Therein, $\overset{\prime\prime}{\mathbf{x}}_\alpha$ is the acceleration of φ^α, "div $(\,\cdot\,)$" is the divergence operator corresponding to "grad $(\,\cdot\,)$", $\mathbf{T}^\alpha$ is the *Cauchy* stress tensor and $\mathbf{g}$ the overall gravity. Concerning the energy balance, ψ^α is the free *Helmholtz* energy, η^α is the entropy, $\mathbf{L}_\alpha = \operatorname{grad} \overset{\prime}{\mathbf{x}}_\alpha$ is the spatial velocity gradient, $\mathbf{q}^\alpha$ is the heat influx vector and r is the radiation. Furthermore, $\hat{\rho}^\alpha$, $\hat{\mathbf{p}}^\alpha$ as well as $\hat{\varepsilon}^\alpha$ are the mass, the direct linear momentum and the direct energy productions satisfying the following constraints:

$$\begin{aligned}
0 &= \sum_\alpha \hat{\rho}^\alpha , \quad \mathbf{0} = \sum_\alpha \hat{\mathbf{p}}^\alpha + \hat{\rho}^\alpha \overset{\prime}{\mathbf{x}}_\alpha , \\
0 &= \sum_\alpha \hat{\varepsilon}^\alpha + \hat{\mathbf{p}}^\alpha \cdot \overset{\prime}{\mathbf{x}}_\alpha + \hat{\rho}^\alpha (\varepsilon^\alpha + \tfrac{1}{2} \overset{\prime}{\mathbf{x}}_\alpha \cdot \overset{\prime}{\mathbf{x}}_\alpha) .
\end{aligned} \tag{5}$$

The deformation of the solid matrix is given within a *Lagrange*an description using the solid displacement vector $\mathbf{u}_S = \mathbf{x} - \mathbf{X}_S$ as the primary kinematic variable, whereas the fluid constituents are described in a modified *Euler*ian setting by use of the seepage velocities $\mathbf{w}_\beta = \overset{\prime}{\mathbf{x}}_\beta - \overset{\prime}{\mathbf{x}}_S$ relating the fluid motions to the deforming skeleton material. Furthermore, one defines the diffusion velocities $\mathbf{d}_{\gamma G} = \overset{\prime}{\mathbf{x}}_\gamma - \overset{\prime}{\mathbf{x}}_G$ of the gas components within the overall gas phase. The temperature θ of the overall mixture and the effective pressure p^{LR} of the pore liquid as well as the partial pressures p_G^γ of the air and the water vapour in the gas phase have to be considered as additional independent fields. The resulting system of equations for the determination of these primary variables consists of the mixture energy balance, the momentum balances under quasi-static conditions ($\overset{\prime\prime}{\mathbf{x}}_\alpha \equiv \mathbf{0}$) and the mass balances of the fluid constituents.

3 Constitutive Modelling

Phase transition processes of water like evaporation and condensation can be described within the TPM using the respective mass production terms. This means in the context of the underlying triphasic material model that

$$\hat{\rho}^S = 0\,, \quad \hat{\rho}^A = 0\,, \quad \hat{\rho}^L + \hat{\rho}^V = 0\,. \tag{6}$$

Furthermore, considering the gas components, the direct momentum production $\hat{\mathbf{p}}^\gamma$ can be additively split in a part $\hat{\mathbf{p}}^\gamma_{\text{int}}$ describing the interaction with the other gas components, whereas the second part $\hat{\mathbf{p}}^\gamma_{\text{ext}}$ describes the interaction with the solid and liquid phases:

$$\hat{\mathbf{p}}^\gamma = \hat{\mathbf{p}}^\gamma_{\text{int}} + \hat{\mathbf{p}}^\gamma_{\text{ext}}\,, \quad \sum_\gamma \hat{\mathbf{p}}^\gamma_{\text{ext}} = \hat{\mathbf{p}}^G\,, \quad \sum_\gamma \hat{\mathbf{p}}^\gamma_{\text{int}} = \mathbf{0}\,. \tag{7}$$

Concerning these assumptions, constitutive equations are required for

$$\psi^\alpha\,,\ \eta^\alpha\,,\ \mathbf{T}^\alpha\,,\ \hat{\mathbf{p}}^\beta\,,\ \hat{\mathbf{p}}^\gamma_{\text{int}}\,,\ s^L\,,\ \mathbf{q}^\alpha\,,\ \rho^{SR}\,. \tag{8}$$

By an evaluation of the entropy inequality, one finds the constraints that the entropies η^S and η^L of the solid and liquid phases, the *Cauchy* stress $\mathbf{T}^\alpha$ as well as the momentum production $\hat{\mathbf{p}}^\beta$ of the fluid phases should consist of two parts, where the first is governed by the corresponding pore pressure, while the second, the so-called "extra term" $(\,\cdot\,)_E$, results from the solid deformation or from the pore-fluid flow, respectively, cf. Ehlers (1993), Ghadiani (2005):

$$\begin{aligned}
\eta^S &= \frac{1}{(\rho^{SR})^2}\frac{\partial\rho^{SR}}{\partial\theta^S}\,p^{FR} + \eta^S_E\,, \quad \eta^L = \frac{1}{(\rho^{LR})^2}\frac{\partial\rho^{LR}}{\partial\theta^L}\,p^{LR} + \frac{\theta^L}{\theta^S}\,\eta^L_E\,,\\
\mathbf{T}^S &= -n^S p^{FR}\,\mathbf{I} + \mathbf{T}^S_E\,, \quad \mathbf{T}^\beta = -n^\beta p^{\beta R}\,\mathbf{I}\,, \quad \mathbf{T}^\gamma = -n^G p^\gamma_G\,\mathbf{I}\,,\\
\hat{\mathbf{p}}^\beta &= p^{\beta R}\,\operatorname{grad} n^\beta + \hat{\mathbf{p}}^\beta_E\,,
\end{aligned} \tag{9}$$

where $p^{FR} = s^L p^{LR} + s^G p^{GR}$ is the pore fluid pressure and p^{LR} and p^{GR} the effective pressures of the liquid and the overall gas phase. Furthermore, the extra stresses $\mathbf{T}_E^\beta$ and $\mathbf{T}_E^\gamma$ are neglected in comparison with the interaction terms $\hat{\mathbf{p}}_E^\beta$, which can be motivated by a dimensional analysis.

Following the arguments by Ghadiani (2005), the material deformation gradient $\mathbf{F}_S$ of the solid skeleton can be multiplicatively decomposed into a mechanical and a thermal part, i.e., $\mathbf{F}_S = \mathbf{F}_{S_m}\mathbf{F}_{S_\theta}$. Further examination under consideration of the mass balance $(4)_1$ yields that the material density of the solid skeleton is purely governed by the thermal part, whereas the volume fraction n^S of the solid phase is purely governed by the mechanical part of the deformation gradient (Ghadiani 2005). In the present contribution, the porous material is considered within the frame of a geometrically linear theory. Thus, the geometrically linearized total strain tensor $\boldsymbol{\varepsilon}_S$ is introduced and additively split into elastic and plastic parts: $\boldsymbol{\varepsilon}_S = \boldsymbol{\varepsilon}_{Se} + \boldsymbol{\varepsilon}_{Sp}$. Therefore, with a given constitutive relation for the material density of the solid skeleton, the volume fraction of the solid matrix can be computed directly based on the mass balance:

$$\rho^{SR} = \rho_{0S}^{SR}\,[1 - 3\alpha^S(\theta - \theta_0)]\,, \quad n^S = n_{0S}^S\,[1 - \operatorname{div}\mathbf{u}_S + 3\alpha^S(\theta - \theta_0)]\,. \quad (10)$$

Therein, α^S is the thermal expansion coefficient of the solid skeleton, and θ_0 is a reference temperature, cf. Ghadiani (2005). One underlying assumption for the free *Helmholtz* energies of the individual constituents is the definition of the specific heat C_V^α. Furthermore, the free *Helmholtz* energy of the solid skeleton is governed by a thermal extended *Hooke*an elasticity law, whereas the free *Helmholtz* energies of the gas components are governed by the ideal gas law $\rho_G^\gamma = (p_0 + p_G^\gamma)/\bar{R}^\gamma\theta$, where $\bar{R}^\gamma$ is the specific gas constant. Therefore:

$$\begin{aligned} \rho_{0S}^S\psi^S &= \mu^S\,\boldsymbol{\varepsilon}_{Se}\cdot\boldsymbol{\varepsilon}_{Se} + \tfrac{1}{2}\lambda^S(\boldsymbol{\varepsilon}_{Se}\cdot\mathbf{I})^2 + m^S\Delta\theta\boldsymbol{\varepsilon}_{Se}\cdot\mathbf{I} - \rho_{0S}^S C_V^S(\theta\ln\frac{\theta}{\theta_0} - \Delta\theta)\,, \\ \psi^L &= -C_V^L(\theta\ln\theta - \theta)\,, \quad \psi^\gamma = \bar{R}^\gamma\ln\rho_G^\gamma\theta - C_V^\gamma(\theta\ln\theta - \theta)\,. \end{aligned} \quad (11)$$

Therein, μ and λ are the *Lamé* constants, and $m^S = -(2\mu^S + 3\lambda^S)\alpha^S$ is the stress-temperature modulus of the solid skeleton. The extra term of the entropies of the solid and liquid phases as well as the entropy of the air and the water vapour can be computed via

$$\eta_E^S = -\frac{\partial\psi^S}{\partial\theta}\,, \quad \eta_E^L = -\frac{\partial\psi^L}{\partial\theta}\,, \quad \eta^\gamma = -\frac{\partial\psi^\gamma}{\partial\theta}\,. \quad (12)$$

Regarding the thermodynamical transport properties of the porous material, the heat influx vectors are given via $\mathbf{q}^\alpha = -\mathbf{H}^\alpha\operatorname{grad}\theta$, where $\mathbf{H}^\alpha$ is the heat conduction tensor of the single constituents. Proceeding from $(11)_1$ for the free *Helmholtz* energy of the solid skeleton, the solid extra stress can be computed via $\mathbf{T}_E^S = \rho_{0S}^S\partial\psi^S/\partial\boldsymbol{\varepsilon}_{Se}$ yielding a thermoelastic law of *Hooke*an type. Considering the underlying concept of viscoplasticity using a single surface yield criterion and a non-associated flow rule, the interested reader is referred to Ehlers et al. (2004).

Relations for the extra term of the linear momentum production $\hat{\mathbf{p}}^\beta_E$ of the liquid and the gas phases as well as for the linear momentum production $\hat{\mathbf{p}}^\gamma_{\text{int}}$ of the gas components can be given via

$$
\begin{aligned}
\hat{\mathbf{p}}^\beta_E &= -(n^\beta)^2 \gamma^{\beta R} (\kappa^\beta_r \mathbf{K}^\beta)^{-1} \mathbf{w}_\beta \,, \\
\hat{\mathbf{p}}^\gamma_{\text{int}} &= -(n^G)^2 (p^{GR})^2 (p_0 \mathbf{D})^{-1} (\theta/\theta_{\text{ref}})^z \mathbf{d}_{\gamma G} \,.
\end{aligned}
\tag{13}
$$

Therein, $\gamma^{\beta R}$ is the specific weight, κ^β_r is the relative permeability factor chosen according to van Genuchten (1980), $\mathbf{K}^\beta$ is the *Darcy* permeability tensor, $\mathbf{D}$ is the diffusion tensor, θ_{ref} is the referential temperature and z an additional material parameter. For a closer look at this topic, the interested reader is referred to Class (2001) and Ehlers et al. (2004). Furthermore, the saturation s^β is computed from the capillary pressure $p^C = p^{GR} - p^{LR}$, also using a formulation following van Genuchten (1980).

As was mentioned above, the pore gas is considered as a mixture of ideal gases saturated with water vapour. Therefore, the following relations hold:

$$
p^{GR} + p_0 = \sum_\gamma (p^\gamma_G + p_0) \,, \quad x^\gamma_m = \frac{p^\gamma_G + p_0}{p^{GR} + p_0} \,. \tag{14}
$$

Therein, x^γ_m is the molar fraction of φ^γ, whereas the partial pressure of the water vapour is given via a relation following *Antoine*,

$$
p^V_G + p_0 = 10^{[A - B/(\theta + C)]} \cdot 133.4 \quad [\text{Pa}] \,, \tag{15}
$$

where A, B and C are additional material parameters and p_0 is the ambient pressure, cf. Class (2001). Therefore, the partial pressure of the water vapour in the gas phase loses the status as a primary variable.

Inserting these assumptions into the fluid momentum balance relations leads to *Darcy*-type equations for the filter velocities and relations for the diffusion velocities corresponding to *Fick*'s law:

$$
\begin{aligned}
n^\beta \mathbf{w}_\beta &= -\frac{\kappa^\beta_r \mathbf{K}^\beta}{\gamma^{\beta R}} \, [\operatorname{grad} p^{\beta R} - \rho^{\beta R} \mathbf{b}] \,, \\
n^G \mathbf{d}_{\gamma G} &= -\frac{p_0 \mathbf{D}}{p^{GR}} \, (\theta/\theta_{\text{ref}})^z \operatorname{grad} x^\gamma_m \,.
\end{aligned}
\tag{16}
$$

Therefore, the filter and diffusion velocities lose their status as primary variables of the triphasic model. The resulting system of equations for the determination of the remaining primary variables consists of the mixture energy balance, the mixture momentum balance and the mass balances of the water ($\varphi^W = \varphi^L + \varphi^V$) and the air component. Further information can be obtained from Lewis and Schrefler (1998), Ehlers et al. (2004), Sanavia et al. (2005).

4 Numerical Example

In the frame of the finite element method, the numerical treatment is based on the weak formulations of the governing equations. In this contribution, an injection process of heated air in a partially saturated soil is considered as numerical example. This process is simulated setting a respective heat and mass fluxes as *Neumann* boundary conditions for the pore air in the injection area (Fig. 1 (**a**)). In Figure 1 (**b**), the distribution of the liquid saturation s^L for a computation with and without mass exchanges is given, whereas in Fig. 2, the temperature distribution of the whole domain (**a**) and a temperature comparison between the computation with and without mass exchanges in a vertical cut through the right boundary of the domain is presented (**b**). The results show that additional pore gas is produced if mass exchange effects are considered. Furthermore, the model is capable to describe the phenomena of latent heat during a phase transformation in a correct way. More precisely, this means that a certain energy is necessary to transfer water from the liquid to the gaseous state without a temperature increase. In this example, where the heat flux over the boundary is constant, this effect leads to a lower temperature at the boundary if mass exchanges are considered.

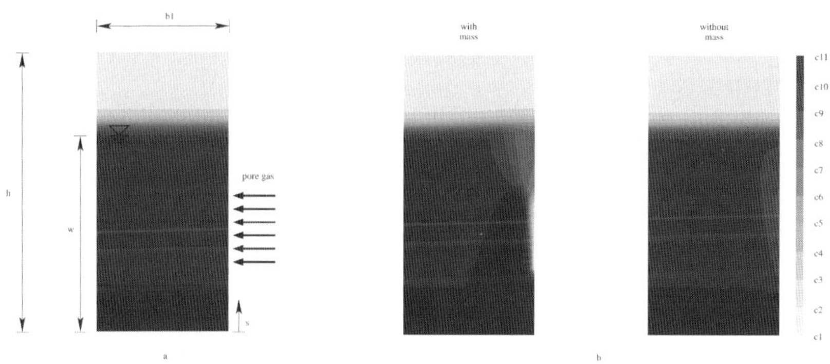

Fig. 1. Sketch of the problem (**a**) and distribution of the liquid saturation s^L (**b**)

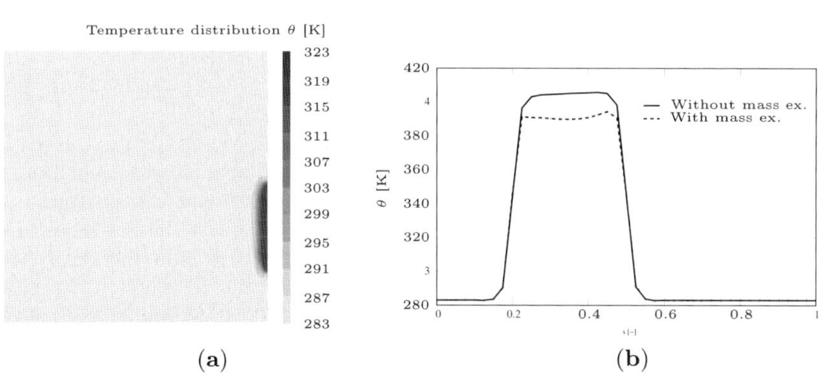

Fig. 2. Distribution of the temperature θ at $t = 22\,\text{h}$

5 Conclusion

In the present contribution, a triphasic model for the description of partially saturated soil under non-isothermal conditions has been considered on the basis of the TPM. In particular, a thermodynamically consistent framework of an elasto-viscoplastic soil skeleton saturated by liquid and gaseous water (water vapour) as well as air has been presented. The gas phase has been assumed as a water-vapour-saturated mixture of the ideal gases air and water vapour. On the numerical side, the triphasic model was implemented in the finite element tool `PANDAS` such that a simulation of an injection process of heated air in a water-saturated porous material could be carried out. This example clearly pointed out that the presented model is qualified to simulate phase transition processes between the pore fluids in a physically correct way.

Acknowledgements

The authors would like to thank the Deutsche Forschungsgemeinschaft (DFG) for the financial support in the frame of the Research Unit 444 *Mechanik teilgesättigter Böden* under grant No. Eh107/14.

References

Class H (2001) Theorie und numerische Modellierung nichtisothermer Mehrphasenprozesse in NAPL-kontaminierten porösen Medien. Dissertation, Mitteilungsheft Nr. 105 aus dem Institut für Wasserbau, Universität Stuttgart

Ehlers W (1993) Constitutive equations for granular materials in geomechanical context. In: Hutter K (ed) Continuum Mechanics in Environmental Sciences and Geophysics, CISM Courses and Lectures No 337. Springer-Verlag, Wien, pp 313–402

Ehlers W (2002) Foundations of multiphasic and porous materials. In: Ehlers W, Bluhm J (eds) Porous Media: Theory, Experiments and Numerical Applications. Springer-Verlag, Berlin, pp 3–86

Ehlers W, Graf T, Ammann M (2004) Deformation and localization analysis of partially saturated soil, Comp Meth Appl Mech Eng 193:2885–2910

van Genuchten MT (1980) A closed-form equation for predicting the hydraulic conductivity of unsaturated soils, Soil Sci Soc Am J 44:892–898

Ghadiani S (2005) A Multiphasic Continuum Mechanical Model for the Design Investigation of an Effusion-cooled Rocket Trust Chamber. Dissertation, Bericht Nr. II–13 aus dem Institut für Mechanik (Bauwesen), Universität Stuttgart

Lewis RW, Schrefler BA (1998) The Finite Element Method in the Static and Dynamic Deformation and Consolidation of Porous Media, 2nd edition. Wiley, Chichester

Sanavia L, Pesavento F, Schrefler BA (2005) Strain localisation simulation in non-isothermal multiphase geomaterials. In: Schanz T (ed) Proceedings of the ISSMGE International Conference from Experimental Evidence towards Numerical Modelling of Unsaturated Soils. Springer-Verlag, Berlin, pp 29–40

A Stress-Strain Framework for Modelling the Behaviour of Unsaturated Soils under Non-Isothermal Conditions

Bertrand François and Lyesse Laloui

Soil Mechanics Laboratory, Ecole Polytechnique Fédérale de Lausanne, Station 18, CH-1015 Lausanne, Switzerland
bertrand.francois@epfl.ch, lyesse.laloui@epfl.ch

Summary. This paper addresses a new, unified thermo-mechanical constitutive model for unsaturated soils through a coupled study. In the context of elasto-plasticity and the critical state theory, the model uses the concepts of multi-mechanism and bounding surface theory. This advanced constitutive approach involves thermo-plasticity of saturated and unsaturated soils. Bishop's effective stress framework is adopted to represent the stress state in the soil. This stress is linked to the water retention curve, which is represented by an elasto-plastic model. Attention is focused particularly on the coupling relations inferred from this unified thermo-hydro-mechanical (THM) study. Finally, the theoretical aspects of the paper are supported by comparisons between numerical simulations and experimental results.

Key words: unsaturated soils, thermo-mechanics, constitutive modelling, elasto-plasticity, multi-mechanism

Introduction

Research interest in the thermo-mechanical behaviour of unsaturated soils is growing as a result of an increasing number of geomechanical problems involving both thermal and capillary effects. In the field of environmental geomechanics, several relevant applications, such as feasibility studies of waste disposal, petroleum extraction or geothermal structures, require an accurate knowledge of the thermo-mechanical behaviour of unsaturated soils (Vulliet et al. 2002). Until now, these two loading cases have often been considered independently.

This paper addresses a new, unified thermo-mechanical constitutive model for unsaturated soils, named ACMEG-TS ("Advanced Constitutive Model in Environmental Geomechanics – Thermal and Suction effects"). This constitutive model is based on thermo-hydro-mechanical (THM) experimental data and attention is focused on the coupling relations linking the thermal, hydric and mechanical soil response. After presenting the stress framework used here, the different experimental data for THM soil behaviour are addressed. The constitutive equations are then introduced. Finally, ACMEG-TS is validated for several typical THM paths.

Stress Framework

An unsaturated state within the soil means that a second fluid phase appears in the inter-particular spaces. The difference in pressure between these two phases (air and water) leads to a new stress variable, suction: $s = u_a - u_w$ which modifies the internal stress state in the soil. Bishop's effective stress is very useful for continuously describing the three phase nature of unsaturated soils (Bishop 1959, Khalili et al. 2005):

$$\sigma'_{ij} = (\sigma_{ij} - u_a \delta_{ij}) + \chi (u_a - u_w) \delta_{ij} \tag{1}$$

where σ_{ij} is the total external stress, u_a and u_w are the air and water pore pressures, respectively, χ the effective stress parameter and δ_{ij} the Kroenecker's symbol. ACMEG-TS uses the generalised Bishop's effective stress, where χ is equal to the degree of saturation S_r, as suggested in the initial work of Bishop (1959) and implemented by Schrefler (1984). Equation (1) requires hydric information (represented by S_r) in parallel with the stress state within the soil. The degree of saturation is related to suction through the retention curve. It is shown that this S_r versus s curve depends on temperature (Salager et al. 2006), on the dry density of the soils (Salager et al. 2007) and on the hydric paths followed (wetting or drying). Thus, the effective stress depends on the thermal, hydric and mechanical histories of the material. Therefore, this single stress approach converts a complex, multi-phase and multi-stress medium in which multi-physics processes occur into a single mechanical state through several coupling equations.

THM Experimental Data

Under normally consolidated conditions (NC), clayey soil contracts when it is heated and a significant part of this deformation is irreversible upon cooling (Laloui 2001). This behaviour over the whole cycle is representative of thermal hardening. Another important non-isothermal behaviour is the fact that the apparent preconsolidation pressure decreases with increasing temperature while the isotropic compressibilities do not seem to be significantly affected by temperature changes (Laloui and Cekerevac 2003). In addition to these thermal effects on the isotropic soil response, hydric conditions partially govern the mechanical response of the soil. Indeed, it is shown that the apparent preconsolidation pressure increases when suction increases for suctions greater than the air entry suction, s_{e0}, and is uninfluenced by suctions lower than s_{e0} (Geiser et al. 2006).

The thermal, as well as the capillary, effects on soil strength seem to be negligible. Cekerevac and Laloui (2004) summarized some experimental results which tend to confirm that the friction angle at critical state can either slightly increase or decrease with temperature. Moreover, in the generalized Bishop's effective stress reference, the friction angle appears to be independent of the amount of suction (Khalili et al. 2004).

ACMEG-TS Constitutive Framework

Mechanical Framework

The basic concept of the ACMEG-TS model is to consider that the thermal, as well as the hydric, loads exclusively involve volumetric effects in soils. These considerations are introduced in an (THM) elasto-plastic framework where each THM loading (external mechanical load, σ, temperature, T, and suction, s) may imply reversible and irreversible changes in the state of the material. Within this elasto-plastic framework, the total strain rate tensor, $\dot{\varepsilon}$, due to THM loading is decomposed into non-linear, thermo-elastic, $\dot{\varepsilon}^e$, and thermo-plastic, $\dot{\varepsilon}^p$, components. Due to the isotropic effect of the temperature and the hydric changes in the soil, the plastic mechanism of the material is induced by two coupled hardening processes: an isotropic one which may be activated by any mechanical, thermal or hydric loads and a deviatoric mechanism acting only under a mechanical loading having a deviatoric component. Both mechanisms may induce volumetric plastic strain (Laloui et al. 2005). Therefore the total volumetric plastic strain rate $\dot{\varepsilon}_v^p$ is the coupling variable linking the two hardening processes and is given by:

$$\dot{\varepsilon}_v^p = \dot{\varepsilon}_{v,\text{iso}}^p + \dot{\varepsilon}_{v,\text{dev}}^p \tag{2}$$

where $\dot{\varepsilon}_{v,\text{iso}}^p$ and $\dot{\varepsilon}_{v,\text{dev}}^p$ are the volumetric plastic strain rate induced by the isotropic and the deviatoric mechanisms, respectively. The yield functions of the two mechanical, thermo-plastic mechanisms have the following expressions (Fig. 1):

$$f_{\text{iso}} = p' - \sigma_c' r_{\text{iso}} = 0\,; \quad f_{\text{dev}} = q - Mp'\left(1 - b\,\text{Log}\,\frac{dp'}{\sigma_c'}\right) r_{\text{dev}} = 0 \tag{3}$$

where the preconsolidation pressure, σ_c', depends on the volumetric plastic strain, ε_v^p, temperature, T, and suction, s (Laloui and Nuth 2005):

$$\sigma_c' = \begin{cases} \sigma_{c0}' \exp(\beta\varepsilon_v^p)\left\{1 - \gamma_T \log\left(\frac{T}{T_0}\right) + \gamma_s \log\left(\frac{s}{s_{e0}}\right)\right\} & \text{if } s > s_{e0}\,, \\ \sigma_{c0}' \exp(\beta\varepsilon_v^p)\left\{1 - \gamma_T \log\left(\frac{T}{T_0}\right)\right\} & \text{if } s \le s_{e0}\,. \end{cases} \tag{4}$$

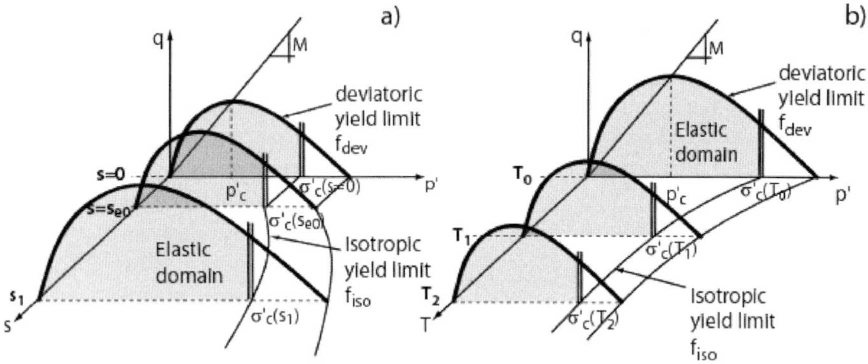

Fig. 1. Hydric (**a**) and thermal (**b**) effects on the THM yield limits

r_{iso} and r_{dev} are the degree of mobilization of the isotropic and the deviatoric mechanisms and are hyperbolic functions of the plastic volumetric strain induced by the isotropic and the deviatoric mechanisms, respectively. They vary from r^e, which defines the elastic domain ($0 < r^e < 1$), to 1 at a completely plastic state (Hujeux 1979).

γ_T and γ_s are the material parameters required to quantify the effect of temperature and suction, respectively, on the apparent preconsolidation pressure. β is the plastic compressibility modulus (the slope of the linear function $\varepsilon_v^p - \log \sigma_c'$). s_{e0} is the air entry suction which depends on the temperature and plastic volumetric strain, as expressed in equation (6).

Hydric Framework

In terms of hydric response, the desaturation process is also seen as a yielding phenomenon. As long as the soil is drying, suction increases and the degree of saturation, S_r, tends to decrease mainly when the air entry suction, s_{e0}, is reached. In this way, s_{e0} can be seen as a hydric yield limit with the following yield function:

$$f_{\text{hyd}} = s - s_e = 0 \tag{5}$$

where the hydric limit, s_e, depends on the degree of saturation, the volumetric plastic strain and the temperature (Fig. 2):

$$s_e = s_{e0} \exp(-\beta_h \Delta S_r) \left\{ 1 - \theta_T \log\left(\frac{T}{T_0}\right) - \theta_e \log(1 - \varepsilon_v^p) \right\} . \tag{6}$$

θ_T and θ_e are the material parameters required to quantify the effect of temperature and volumetric plastic strain on the evolution of the air entry suction, respectively. β_h is the slope of the linear function $S_r - \log s$ when the air entry suction is overcome. For the wetting process, the same elasto-plastic concept can also be applied by considering the hydric hysteresis through an additional material parameter, s_{hys}, quantifying the hydric irreversibility (Fig. 2).

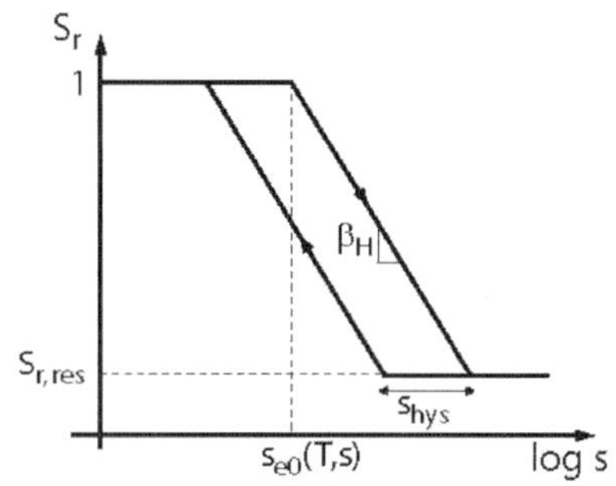

Fig. 2. The air entry suction as the key parameter in the water retention curve

Numerical Simulations

A set of oedometric compression tests were carried on a remoulded sandy silt at different temperatures and suctions (François et al. 2007). The results of these experiments clearly show the thermal and hydric effects on the apparent preconsolidation pressure. The comparison between numerical simulations and experimental results show the ability of ACMEG-TS to represent such THM coupling (Fig. 3). For this simulation, the three key parameters are $\{\gamma_T;\ \gamma_s;\ \beta\} = \{0.8;\ 0.28;\ 37\}$. Moreover, the model is able to reproduce the typical thermal hardening observed along a temperature cycle (Baldi et al. 1991) (Fig. 4).

Conclusions

A highly-coupled THM model (ACMEG-TS) has been presented. It is based on two interrelated constitutive schemes, the mechanical and the hydric one.

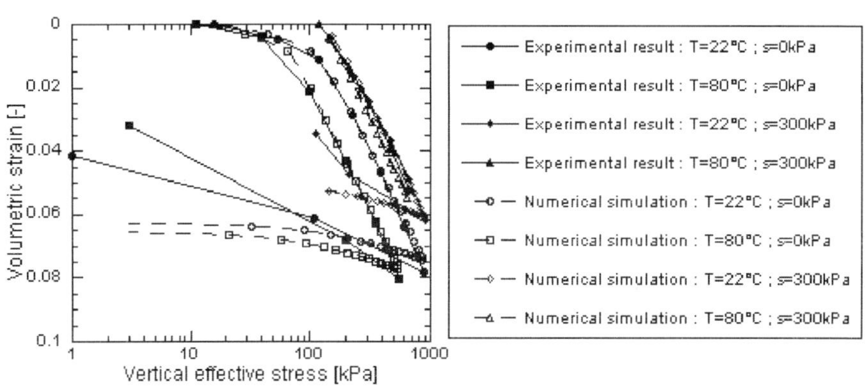

Fig. 3. Numerical simulations of THM oedometric compression tests on a sandy silt

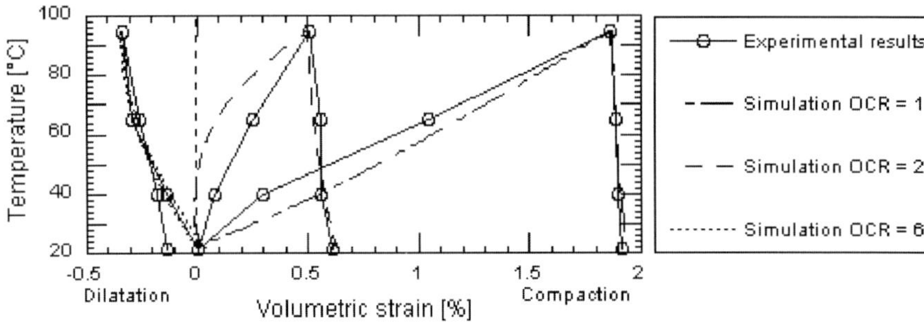

Fig. 4. Numerical simulations of the mechanical response of Boom Clay along a thermal cycle

The hydric condition influences the isotropic mechanical behaviour via its effect on the apparent preconsolidation pressure. Inversely, the mechanical state may have an effect on the hydric response by modifying the water retention curve (via the evolution of the air entry suction). Moreover, these two-way couplings depend on temperature. Therefore, this THM constitutive model considers the three loading cases (external load, suction and temperature) as three interconnected phenomena.

Acknowledgements

This work was partly funded by the Swiss State Secretariat for Education and Research SER, Grant OFES C04.0021.

References

Baldi G, Hueckel T, Peano A, Pellegrini R (1991) Developments in modelling of thermo-hydro-mechanical behaviour of Boom clay and clay-based buffer materials (Vol 1 and 2). EUR 13365/1 and 13365/2, Luxembourg

Bishop AW (1959) The principle of effective stress, Teknisk Ukeblad 39:859–863

Cekerevac C, Laloui L (2004) Experimental study of thermal effects on the mechanical behaviour of clays, Int J Num Anal Meth in Geomechanics 28:209–228

François B, Salager S, El Youssoufi MS, Ubals Picanyol D, Laloui L, Saix C (2007) Compression tests on a sandy silt at different suction and temperature level. In: GeoDenver07 ASCE Conference (Submitted)

Geiser F, Laloui L, Vulliet L (2006) Elasto-plasticity of unsaturated soils: laboratory test results on a remoulded silt, Soils and Foundations Journal 46(5) (in press)

Hujeux J (1979) Calcul numérique de problèmes de consolidation élastoplastique. PhD Thesis, Ecole Centrale Paris

Khalili N, Witt R, Laloui L, Vulliet L, Koliji A (2005) Effective stress in double porous media with two immiscible fluids, Geophysical Research Letters 32(15):L15309

Khalili N, Geiser F, Blight GE (2004) Effective Stress in Unsaturated Soils: Review with New Evidence, Int J Geomech 4(2):115–126

Laloui L (2001) Thermo-mechanical behaviour of soils, Revue Française de Génie Civil 5:809–843

Laloui L, Cekerevac C (2003) Thermo-plasticity of clays: an isotropic yield mechanism, Computers and Geotechnics 30:649–660

Laloui L, Cekerevac C, François B (2005) Constitutive modelling of the thermo-plastic behaviour of soils, Revue Européenne de Génie Civil 9:635-650

Laloui L, Nuth M (2005) An introduction to the constitutive modelling of unsaturated soils, Revue Européenne de Génie Civil 9(5–6):651–670

Salager S, El Youssoufi MS, Saix C (2007) Experimental study of the water retention curve as a function of void ratio. In: GeoDenver07 ASCE Conference (Submitted)

Salager S, Jamin F, El Youssoufi MS, Saix C (2006) Influence de la température sur la courbe de rétention d'eau de milieux poreux. C. R. Mécanique 334:393–398

Schrefler BA (1984) The finite element method in soil consolidation (with applications to surface subsidence). Ph.D. Thesis, University College of Swansea, C/Ph/76/84

Vulliet L, Laloui L, Schrefler BA (2002) Environmental Geomechanics. EPFL-Press, 423 pp

Numerical Simulation of Multiphase Multicomponent Processes Considering Structural Alterations of Porous Media – a Phenomenological Model

Sandra Freiboth, Holger Class, and Rainer Helmig

Institute of Hydraulic Engineering, Department of Hydromechanics and Modeling of Hydrosystems, Pfaffenwaldring 61, 70550 Stuttgart, Germany
Sandra.Freiboth@iws.uni-stuttgart.de, Holger.Class@iws.uni-stuttgart.de, Rainer.Helmig@iws.uni-stuttgart.de

Summary. The numerical simulation of flow and transport processes is frequently applied to environmental, technical and even medical problems. In many cases, a deformation of the porous medium occurs which cannot be neglected. Therefore, we present a non-isothermal multiphase multicomponent flow and transport model which also allows for structural alterations of the porous medium phenomenologically. This means that the model accounts for the effects of the structural changes by adapting the hydraulic properties using constitutive relationships.

Key words: multiphase modelling, structural alterations, swelling, shrinkage, hydraulic properties

1 Introduction

The numerical simulation of multiphase multicomponent flow and transport processes in porous media is a valuable tool for a variety of environmental, technical and also medical problems. For many applications in these fields, the porous medium cannot be assumed to be rigid. Such deformable porous media are, for example, cohesive soils, diapers and the mammalian brain. Multiphase flow and transport processes in cohesive soils are of great interest as regards the long-term stability and integrity of nuclear repositories and municipal waste-disposal sites. The numerical simulation of flow and transport processes in the mammalian brain may be used to optimise the treatment of cerebral tumours.

In both cases, the focus lies on the flow and transport processes, but these processes are strongly influenced by the structural alteration of the porous medium. Thus, we present a model concept for two-phase two-component flow and transport in which the structural alterations are considered phenomenologically: the effects of swelling and shrinking on the flow and transport processes are taken into account. This is achieved by adapting the relevant hydraulic parameters using constitutive relationships for the porosity, the permeability and the capillary pressure.

In the following section, we present our conceptual model for non-isothermal two-phase two-component flow and transport in a rigid porous medium and in the subsequent section the extension for deformable porous media. After that, we demonstrate a desiccation process in a clay liner as an application example. Finally, we give an outlook on our parameter identification strategy.

2 Conceptual Model for a Rigid Porous Medium

The non-isothermal two-phase two-component model for rigid porous media is explained in Helmig (1997), Emmert (1997) and Class et al. (2002) in detail. It is a macroscopic approach, which describes all processes based on a representative elementary volume (REV). To define the system state some variables are needed which quantify the amount of substance, the pressures and the thermal conditions.

The volume occupied by the two fluid phases gas and water is specified by the porosity φ, which is the volume fraction of pore space. The distribution of the fluid phases ($\alpha \in w, g$) in the pore space is quantified by phase saturations S_α which are defined as the ratio of phase volume to pore volume. Thus, all phases sum up to 1:

$$\sum_{\alpha=1}^{2} S_\alpha = S_w + S_g = 1 \,. \tag{1}$$

As we consider mass transfer between the phases (dissolution of air in water and vaporisation of water), the composition of each phase of the components water and air ($\kappa \in w, a$) has to be quantified by mole fractions x_α^κ. By definition, all mole fractions for each phase sum up to 1:

$$\sum_{\kappa=1}^{2} x_\alpha^\kappa = x_\alpha^w + x_\alpha^a = 1 \,. \tag{2}$$

As a result of capillary forces the pressures of both phases might differ. The pressure difference is called capillary pressure:

$$p_c = p_g - p_w \,. \tag{3}$$

On the macroscale, the capillary pressure is a function of water saturation, for which several parameterisations exist (e.g. after Genuchten 1980).

Since we assume local thermal equilibrium between all fluid phases and the solid phase in each REV, only one temperature value T is needed to describe the thermal conditions.

For the unique description of the system state, only three variables are necessary. These so-called primary variables are chosen depending on the phases present in the REV (the so-called phase state). All other – secondary – variables are derived from the above-mentioned constraints and further thermodynamic laws linking the mole fraction of the liquid phase to the partial pressures of the gaseous phase.

In addition to the mass-transfer processes between the phases (vaporisation and condensation; dissolution and degassing) the following processes are considered:

- advective mass and heat transport with the phases due to pressure gradients, phase velocities are calculated according to the extended Darcy's Law,
- diffusive mass and heat transport within the phases due to concentration gradients, diffusive fluxes are given by Fick's Law,
- conductive heat transfer in the entire porous medium due to temperature gradients, heat fluxes are given by Fourier's Law.

To describe the non-isothermal two-phase two-component system mathematically, one mass balance equation for each component (eq. 4) and one energy balance equation for the entire porous medium (eq. 5) are set up

$$\frac{\mathrm{d}(\varphi \sum_\alpha \varrho_{\mathrm{mol}\alpha} x_\alpha^\kappa S_\alpha)}{\mathrm{d}t} - \sum_\alpha \nabla \cdot \left(\frac{k_{r\alpha}}{\mu_\alpha} \varrho_{\mathrm{mol}\alpha} x_\alpha^\kappa \mathsf{K} (\nabla p_\alpha - \varrho_\alpha \mathbf{g}) \right) - \sum_\alpha \nabla \cdot \left(D_{\alpha\,\mathrm{pm}}^\kappa \nabla (\varrho_{\mathrm{mol}\alpha} x_\alpha^\kappa) \right) - q^\kappa = 0, \quad (4)$$

$$\frac{\mathrm{d}\,(\varphi \sum_\alpha \varrho_\alpha u_\alpha^\kappa S_\alpha)}{\mathrm{d}t} + \frac{\mathrm{d}\,((1-\varphi)\varrho_s c_s T)}{\mathrm{d}t} - \nabla \cdot (\lambda_{\mathrm{pm}}^h \nabla T) - \sum_\alpha \nabla \cdot \left(\frac{k_{r\alpha}}{\mu_\alpha} \varrho_\alpha h_\alpha \mathsf{K} (\nabla p_\alpha - \varrho_\alpha \mathbf{g}) \right) - \sum_\kappa \nabla \cdot \left(D_{g\,\mathrm{pm}}^\kappa h_g^\kappa M^\kappa \nabla (\varrho_{\mathrm{mol}g} x_g^\kappa) \right) - q^h = 0\,. \quad (5)$$

In eqs. (4) and (5), $\varrho_{\mathrm{mol}\alpha}$ denotes the molar density of phase α, $k_{r\alpha}$ the relative permeability, μ_α the dynamic viscosity, K the absolute permeability tensor, ϱ_α the density, $\mathbf{g}$ the vector of acceleration due to gravity, $D_{\alpha\,\mathrm{pm}}^\kappa$ the diffusion coefficient for component κ in phase α for the porous medium, q^κ and q^h source and sink terms, u_α the specific internal energy, c_s the specific heat capacity of the soil, λ_{pm}^h the heat conductivity of the porous medium, and h_α^κ the specific enthalpy. The subscript h stands for heat, the superscript s for solid.

This leads to a system of three coupled, non-linear partial differential equations with three unknown state variables.

The two-phase two-component model is realised within the software package MUFTE-UG, made up of the two parts MUFTE and UG. The physical model and the spatial discretisation are implemented in MUFTE (multiphase flow transport and energy model), while UG (unstructured grid) provides the numerical basis such as multigrid data-structures, solvers, and time discretisation.

For spatial discretisation, a mass conservative, finite-volume method (BOX method) is used. For time discretisation, an implicit Euler scheme is applied. The non-linearities are handled with the Newton–Raphson linearisation.

3 Conceptual Model for a Deformable Porous Medium

The two-phase two-component model is extended so that the effects of structural alterations are considered on a phenomenological basis.

Because of the structural changes, the phase saturations do not necessarily determine the system state uniquely. Therefore, the total amount of water has to be chosen as a primary variable.

Water can be bound in the soil matrix of clays due to their layered mineral structure. This inclusion of water involves a deformation of the minerals and consequently of the soil matrix, too (Scheffer and Schachtschabel 1998). For the macroscale model, this phenomenon is considered by an immobilisation/mobilisation term as well as by an increase/decrease of the bulk and pore volume. Although the bulk volume changes are considered, the model geometry (spatial discretisation and volume) is kept constant. It is assumed that this procedure is sufficiently precise for moderate structural alterations.

The mass of immobile water can be described as a saturation function (in a mathematical sense) of the total mass of water (Meißner and Dobrowolsky 2003). The change in bulk volume is quantified by the swelling/shrinking factor ε which is defined as the ratio between the actual bulk volume and the initial (= computational) bulk volume. The swelling/shrinking factor ε also depends on the total mass of water. Generally, the shrinkage process of a cohesive soil can be divided into two parts: the shrinkage phase from the saturated state to the shrinkage limit and the residual shrinkage phase from the shrinkage limit to the completely dry state. For both phases, a linear relationship between the swelling/shrinking factor and the total mass of water can be applied (Brauns et al. 2000, Tariq and Durnford 1993). The real porosity is approximated as a linear function of the swelling/shrinking factor. As the model geometry, i.e. the element volume, is not changed, a computationally effective porosity φ_{eff} is defined:

$$\varphi_{\text{eff}} = \varepsilon \varphi_{\text{real}} \,. \tag{6}$$

Additionally, the relevant hydraulic properties such as permeability and capillary pressure are adapted by constitutive relationships to account for the influence of the structural alterations on the flow and transport processes. Clauser (2003) presents various approaches for describing the permeability as a function of porosity and other soil properties. These approaches require numerous unknown parameters. Therefore, the relationship between permeability and porosity is approximated by a combination of a linear and a potential function specified by the minimum and maximum permeability at minimum and maximum porosity. On the basis of the Leverett J-function, the capillary pressure is scaled depending on a permeability and porosity in relation to the capillary pressure, the permeability and the porosity of a certain reference state.

Furthermore, effects of vapour-pressure lowering above curved surfaces cannot be neglected in fine-grained soils like clays. A relationship between the vapour-pressure in the porous medium and the capillary pressure can be derived from the Kelvin Equation (Atkins 1990).

This model extension is incorporated into the existing two-phase two-component model so that the flow and transport part of the model is calculated in the first step of a time step and subsequently, in the second step of a time step, the part accounting for structural alterations is computed. Thus, in the second step, all parameters describing the structural state and the hydraulic properties are updated, so that mass and energy conservation is guaranteed. This decoupled consideration involves a limitation of time-step size.

This model is adequate for simulating flow and transport process in a shrinking soil up to the point at which cracks occur.

4 Application Example

Similar to the requirements for a mineral sealing liner according to TASi (1993) and TAA (1991), an application example has been set up. The two-dimensional sample domain ($0.1\,\mathrm{m} \times 3.2\,\mathrm{m}$) is divided into two subdomains (see Fig. 1). The upper layer (0.825 m thick) consists of a homogeneous, shrinking clay with a low permeability and a high capillary pressure. The lower subdomain is made up of a homogeneous, non-shrinking sand with a high permeability and a low capillary pressure.

The top boundary is impermeable for water, at the bottom boundary a low water saturation ($S_w = 0.1$) is set. The gas-phase pressure at the top and the bottom is equal to the atmospheric pressure. The temperature is fixed to $T_{\mathrm{top}} = 335\,\mathrm{K}$ at the top and to $T_{\mathrm{bottom}} = 285\,\mathrm{K}$ at the bottom. The lateral boundaries are entirely impermeable for mass and energy.

Initially, the clay layer is fully water saturated, while the sand base is rather dry ($S_w = 0.1$). The gas-phase pressure and the temperature are constant over the whole domain at the beginning (atmospheric pressure and $T = 285\,\mathrm{K}$).

With this set-up, a desiccation process in the clay layer is simulated with the two-phase two-component model, neglecting any structural changes (in the following abbreviated as *2p2cni model*) and with the two-phase two-component model described above accounting for structural alterations

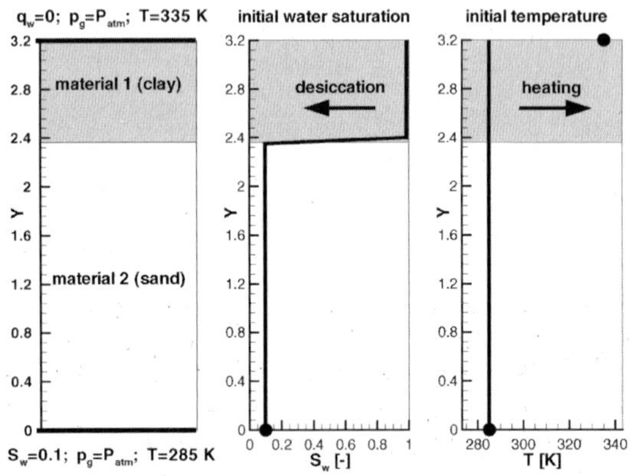

Fig. 1. Application example: domain and boundary conditions (left), initial water saturation (middle) and initial temperature (right)

(*2p2cni-shrink model*). The 2p2cni-shrink model and the 2p2cni model are compared to each other to point out the differences of both models. The simulation results are presented in Fig. 2. It depicts the water saturation S_w, the relative total mass of water $m^w_{\mathrm{tot;rel}}$, the porosity φ, the shrinking factor ε, the permeability K, and the capillary pressure p_c at a certain point in the clay layer over time for both models.

The course of the water saturation obtained with the 2p2cni model is characterised by three phases: an advection-dominated phase at the beginning, an diffusion-dominated phase up to 17 years and an almost stationary phase from 17 years to the end of simulation. Naturally, the relative total mass of water is directly proportional to the water saturation, the permeability is constant and the capillary pressure increases with decreasing water saturation.

In addition to the three above-mentioned phases, the results of the 2p2cni-shrink model clearly show the shrinkage phase (the first six years) and the residual shrinkage phase (afterwards). The shrinkage phase is characterised by a minor decrease of water saturation, but a significant reduction of bulk volume, porosity and permeability as well as an increase of capillary pressure compared to the 2p2cni model. In the residual shrinkage phase, the bulk volume, the porosity and the permeability decrease only slightly. Instead, the water saturation decreases more strongly. In comparison to the 2p2cni model, the desiccation process is slower with the 2p2cni-shrink model. Altogether, the 2p2cni-shrink model simulates the characteristics of a shrinkage process qualitatively correctly.

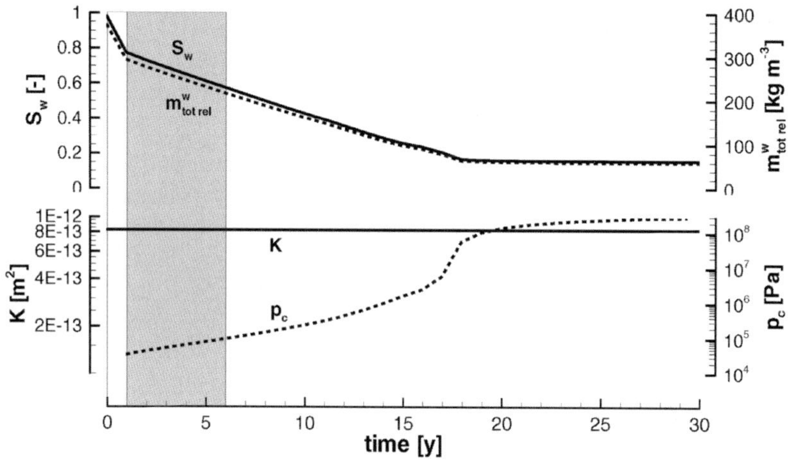

(a) 2p2cni model

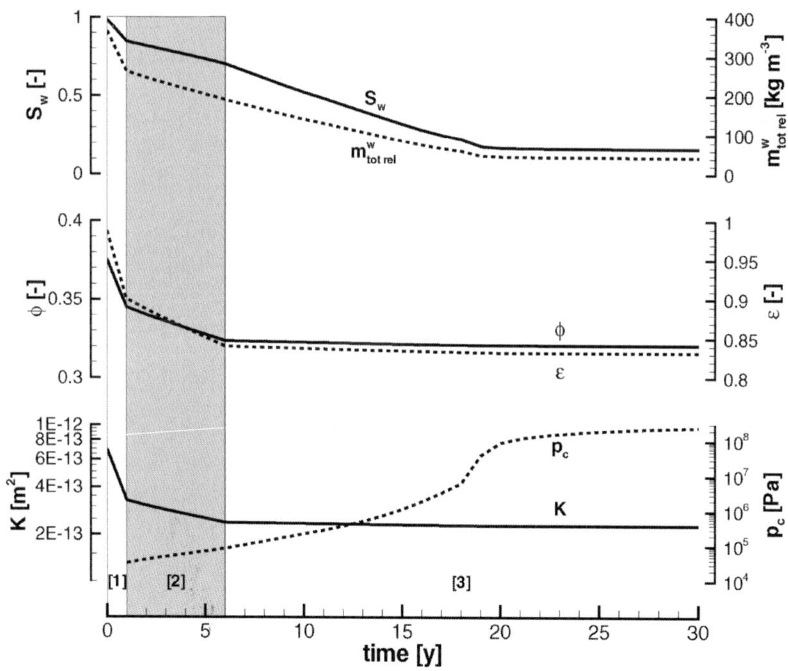

(b) 2p2cni-shrink model

Fig. 2. Comparison of the 2p2cni and the 2p2cni-shrink model: system state and properties at one point in the clay layer ($x = 0.05\,\mathrm{m}$, $y = 2.6\,\mathrm{m}$) over time

5 Outlook onto Parameter Identification

To validate the model, the model results have to be compared to experimental data. However, this raises the problem of parameter identification. As many parameter values have to be determined, a complex parameter-identification strategy is essential. The main parts of this parameter-identification strategy are the experimental design and the application of inverse modelling.

The first step includes the development of a parameter-identification strategy for a non-shrinking soil based on the two-phase two-component model. In the second step, this strategy is transferred and extended to shrinking soils and the extended two-phase two-component model. The main part of both steps is the design of appropriate and efficient experiments.

So far, we have proposed an experimental set-up which will be run by project partners at HTW Dresden this autumn.

References

TAA (TA Abfall) – Technische Anleitung zur Lagerung, chemisch/physikalischen, biologischen Behandlung, Verbrennung und Ablagerung von besonders überwachungsbedürftigen Abfällen, 1991

TASi (TA Siedlungsabfall) – Technische Anleitung zur Verwertung, Behandlung und sonstigen Entsorgung von Siedlungsabfällen, 1993

Atkins PW (1990) Physikalische Chemie. VCH, Weinheim

Brauns J, Schneider H, Gottheil K (2000) Wasserverlust und Schrumpfen bei mineralischen Dichtungen aus geotechnischer Sicht, Geotech 23(1):36–47

Class H, Helmig R, Bastian P (2002) Numerical Simulation of Nonisothermal Multiphase Multicomponent Processes in Porous Media – 1. An Efficient Solution Technique, Adv Water Resour 25(5):533–550

Clauser C (ed) (2003) Numerical Simulation of Reactive Flow in Hot Aquifers – SHEMAT and Processing SHEMAT. Springer, Berlin

Emmert M (1997) Numerische Modellierung nichtisothermer Gas–Wasser Systeme in porösen Medien. Mitteilungsheft, Institut für Wasserbau, Universität Stuttgart

Van Genuchten MT (1980) A Closed-Form Equation for Predicting the Hydraulic Conductivity of Unsaturated Soils, Soil Sci Soc Am J 44:892–898

Helmig R (1997) Multiphase Flow and Transport Processes in the Subsurface: A Contribution to the Modeling of Hydrosystems. Springer, Berlin

Meißner H, Dobrowolsky M (2003) Personal Communication (10.03.2003)

Scheffer F, Schachtschabel P (1998) Lehrbuch der Bodenkunde, 14. edition. Enke, Stuttgart

Tariq A-u-R, Durnford DS (1993) Analytical Volume Change Model for Swelling Clay Soils, Soil Sci Soc Am J 57:1183–1187

Flow and Contaminant Transport Model for Unsaturated Soil

Akbar A. Javadi, Mohammed M. AL-Najjar, and Brian Evans

Department of Engineering, University of Exeter, North Park Road, Harrison building, EX4 4QF, Exeter, UK a.a.javadi@ex.ac.uk, m.m.al-najjar@ex.ac.uk, b.m.evans@ex.ac.uk

Summary. This paper presents the development and validation of a numerical model for simulation of the flow of water and air and contaminant transport through unsaturated soils. The governing differential equations include two mass balance equations for the water phase and air phase together with a balance equation for contaminant transport through the two phases. In the model the nonlinear system of the governing differential equations was solved using a finite element method in the space domain and a finite difference scheme in the time domain. The governing equation of the miscible contaminant transport including advection, dispersion-diffusion and adsorption effects are presented. The model is validated by application to standard experiments on contaminant transport in unsaturated soils. The application of the model to case study is then presented and discussed.

Key words: contaminant transport, unsaturated soil, finite element

1 Introduction

In recent years, interest in understanding the mechanisms and prediction of contaminant transport through soils has dramatically increased because of growing evidence and public concern that the quality of the subsurface environment is being adversely affected by industrial, municipal and agricultural waste. In assessing the environmental impacts of waste discharges, engineers seek to predicate the impact of emission on contaminant concentration in nearby air and water (Nazaroff and Alvarez-Cohen 2001). The high costs, large time scales and lack of control over the boundary conditions have prevented the development of field scale experiments (Hellawell and Sawidou 1994). In this paper, the main governing phenomena of the miscible contaminant transport including advection, mechanical dispersion, molecular diffusion and adsorption are considered. The contaminant transport equation together with the balance equations for flow of water and air are solved numerically using the finite element method, subject to prescribed initial and boundary conditions. The model is used to analyze the transport of a petroleum-based contaminant in a site in the south west of the UK. The results of the model prediction are compared with those measured on the site. It is shown that the developed model is capable of predicting the variations of the contaminant concentration with time with a very high accuracy.

2 Governing Equations of Fluid Flow and Contaminant Transport in Soil

There is a broad range of issues that are of interest in relation to transport of contaminant in soils. The problem becomes more complex when the soil is unsaturated. Unsaturated soil is a multiphase system, because at least two fluid phases are present: water and air. The governing equations that describe fluid flow and contaminant transport in the unsaturated zone will be presented in this section.

2.1 Modeling of Water and Air Flow

The governing differential equation for water flow is based on the conservation of mass of the ground water, leading to (Javadi et al. 2006):

$$c_{ww}\frac{\partial u_w}{\partial t} + c_{wa}\frac{\partial u_a}{\partial t} = \nabla[K_{ww}\nabla u_w] + \nabla[K_{wa}\nabla u_a] + J_w \tag{1}$$

where

$$c_{ww} = c_{fw} + c_{vw}\,, \quad c_{wa} = c_{fa} + c_{va}\,, \quad c_{vw} = nS_aK_{fw}\,, \quad c_{va} = nS_aK_{fa}\,,$$

$$c_{fw} = -n(\rho_w - \rho_v)\frac{\partial S_w}{\partial s}\,, \quad c_{fa} = n(\rho_w - \rho_v)\frac{\partial S_w}{\partial s}\,, \quad K_{ww} = \frac{\rho_w K_w}{\gamma_w} + K_{vw}\rho_w\,,$$

$$K_{wa} = \rho_v K_a + \rho_w K_{va}\,, \quad K_{fa} = \rho_0\frac{\partial h}{\partial \psi}\frac{\partial \psi}{\partial s}\nabla u_a\,, \quad K_{fw} = -\rho_0\frac{\partial h}{\partial \psi}\frac{\partial \psi}{\partial s}\nabla u_w\,,$$

$$K_{vw} = -\frac{D_{atms}V_v n}{\rho_w}K_{fw}\,, \quad K_{va} = -\frac{D_{atms}V_v n}{\rho_w}K_{fa}\,, \quad J_w = \rho_w\nabla(K_w\nabla z)\,,$$

in which n is the porosity of the soil, K_w is the conductivity of water $[\mathrm{L}][\mathrm{T}]^{-1}$, K_a is the conductivity of air $[\mathrm{L}][\mathrm{T}]^{-1}$, S_w is the degree of saturation of water, S_a is the degree of saturation of air, ρ_w is the density of water $[\mathrm{M}][\mathrm{L}]^{-3}$, ρ_v is the density of water vapor $[\mathrm{M}][\mathrm{L}]^{-3}$, ρ_0 is the density of saturated soil vapor $[\mathrm{M}][\mathrm{L}]^{-3}$, s is the soil suction $[\mathrm{M}][\mathrm{L}]^{-1}[\mathrm{T}]^{-2}$, V_v is the mass flow factor, u_w is the pore-water pressure $[\mathrm{M}][\mathrm{L}]^{-1}[\mathrm{T}]^{-2}$, u_a is the pore-air pressure $[\mathrm{M}][\mathrm{L}]^{-1}[\mathrm{T}]^{-2}$, D_{atms} is the molecular diffusivity of vapor through air, γ_w is the unit weight of water $[\mathrm{M}][\mathrm{L}]^{-2}[\mathrm{T}]^{-2}$, ψ is the capillary potential, h is the relative humidity and ∇z is the unit normal oriented downwards in the direction of the force of gravity. The governing differential equation for air flow is based on the conservation of mass of the ground air, leading to (Javadi et al. 2006):

$$c_{aw}\frac{\partial u_w}{\partial t} + c_{aa}\frac{\partial u_a}{\partial t} = \nabla[K_{aw}\nabla u_w] + \nabla(K_{aa}\nabla u_a) + J_a \tag{2}$$

where

$$c_{aw} = c_{aw1} + c_{aw2}\,, \quad c_{aa} = c_{aa1} + c_{aa2}\,, \quad c_{aw1} = -n\rho_{da}(H_a - 1)\frac{\partial S_w}{\partial s}\,,$$

$$c_{aa1} = n\rho_{da}(H_a - 1)\frac{\partial S_w}{\partial s}\,, \quad c_{aw2} = n(S_a + H_a S_w)c_{daw}\,,$$

$$c_{aa2} = n(S_a + H_a S_w)c_{daa}\,, \quad c_{daw} = -\frac{R_v}{R_{da}}K_{fw}\,, \quad c_{daa} = \frac{1}{R_{da}T} - \frac{R_v}{R_{da}}K_{fa}\,,$$

$$K_{aw} = \frac{H_a\rho_{da}}{\gamma_w}K_w\,, \quad K_{aa} = K_a\nabla u_w\,, \quad J_a = H_a\rho_{da}\nabla(K_w\nabla z)\,,$$

in which H_a is the Henry's volumetric coefficient of solubility, ρ_{da} is the density of dry air $[\mathrm{M}][\mathrm{L}]^{-3}$, R_{da} is the specific gas constant for dry air and R_v is the specific gas constant for water vapour.

2.2 Modeling of Contaminant Transport

In porous media, contaminant transport occurs by various processes including advection, mechanical dispersion, diffusion and adsorption. The mass balance equation of contaminant transport can be written as [2]:

$$\frac{\partial(\theta_w c_w)}{\partial t} + \frac{\partial}{\partial t}(\theta_s\rho_s K_d c_w) + \nabla(v_w c_w) - \nabla(\theta_w D_w \nabla c_w) + \lambda_w\theta_w c_w = F^w \tag{3}$$

where θ_w is the volumetric water content, v_w groundwater velocity $[\mathrm{L}][\mathrm{T}]^{-1}$, D_w the coefficients of dispersivity tensor $[\mathrm{L}][\mathrm{T}]^{-1}$, λ_w is the reaction rate for water $[\mathrm{T}]^{-1}$, c_w contaminant concentration K_d is the distribution coefficient and F^w is the source/sink term for water $[\mathrm{M}][\mathrm{L}]^{-3}[\mathrm{T}]^{-1}$. In the case of sorption the equation for the water phase is modified to include a retardation factor. The principal assumption used in deriving a retardation factor is that water is the wetting fluid so that the air phase does not have any contact with the solid phase (Li et al. 1999). Therefore, the equation can be rewritten as:

$$\frac{\partial(R\theta_w c_w)}{\partial t} + \nabla(v_w c_w) - \nabla(\theta_w D_w \nabla c_w) + \lambda_w\theta_w c_w = F^w \tag{4}$$

where the retardation coefficient $R = \left[1 + \frac{\theta_s\rho_s K_d}{\theta_w}\right]$, ρ_s is the density of the solid phase and θ_s is the volumetric content of the solid phase.

3 Case Study

The developed finite element model is applied to a case study involving transport of a petroleum-based contaminant at a site in south west of England, in order to study the potential for contamination from previous commercial use as a fuel filling station and vehicle repair workshop.

3.1 Site Descriptions

The local geology comprises Yeovil Sand beds to 60m depth, with Jurassic limestone immediately to the north. The surface geology of the site includes shallow, fine alluvial deposits containing organic matter, and layers of coarse grained material, weathered limestone with limestone fragments. The site is underlain by a major aquifer and is on the boundary of a fluvial floodplain, having an annual flooding risk of 1%. The garage is adjacent to the junction of two minor roads. It is surrounded by domestic dwellings, with a watercourse approximately 10 m to the north and 2 m below forecourt level, draining to the east. The plot is approximately 20 m by 20 m and consists of a building formerly used as a shop and office, together with two attached workshops with concrete floors, used for repairs and storage. Adjacent to the current office entrance is a store containing two paraffin or light oil tanks, each of 1300 liters capacity. The forecourt is concrete surfaced above the fuel tanks, with a tarmac and gravel access road to the rear. The fuel filling area is directly adjacent to the public pavement and consists of four pumps. Five manhole covers are nearby, at least two of which were found to provide access to fuel storage tanks. Two surface drains are adjacent to the fuel filling.

3.2 Site Observation

A number of inspection covers are present on the site, providing access to fuel tank fillers, pipe manifolds, water supply pipes and two surface drains, with two further drains on the site periphery. Tests showed that one drain adjacent to the fuel pumps discharges directly into an adjacent river which represents an environmental receptor, which means that any spillages from pump operation have a direct pathway to local surface water. Water present beneath some inspection covers showed considerable contamination by heavy oils.

3.3 Sampling

Eight boreholes were installed for monitoring groundwater as shown in Fig. 1. These were all sampled at 0.3 m below groundwater surface to provide a comprehensive sample containing possible dispersed and dissolved fuel compounds. Such contaminants can be expected to show greatest mobility and hence potential for migration off-site. Four monitoring boreholes had previously been installed to three meters depth, adjacent to the storage tanks and pump areas. For the investigation, four additional boreholes were installed as close as possible to the site boundaries. Installation points were selected to surround the site as far as practicable, with emphasis on the north and west boundaries, as observations suggest that groundwater is likely to flow in this direction. The new boreholes, B5–B8, were of a similar design to the original i.e., 50 mm diameter PVC, slotted from 1 m BGL, and were installed to

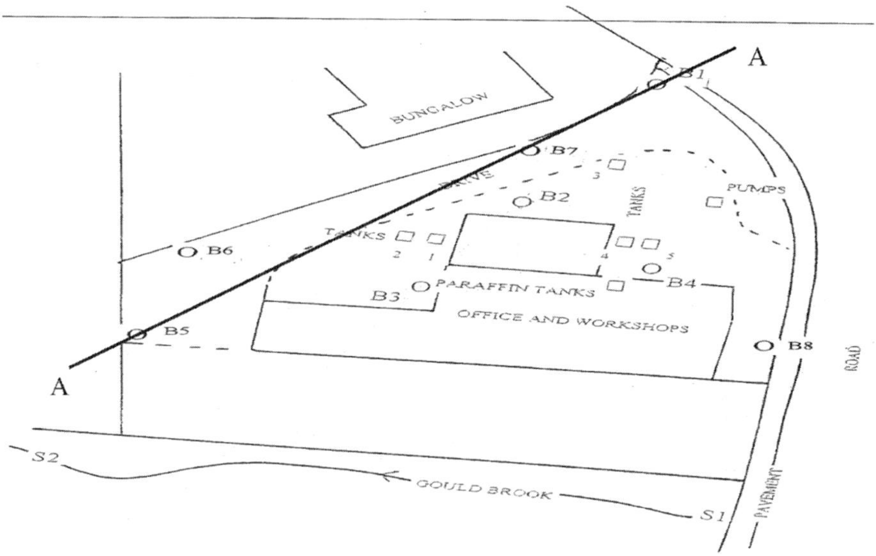

Fig. 1. Plan of the site

a depth of five meters. Groundwater in the boreholes was allowed to equilibrate and was sampled four days after installation. Water samples were taken 0.3 m below groundwater surface to exclude floating product, which may be constrained on the site, and to detect dispersed and dissolved fuel components which are more vulnerable to migration with groundwater. The river was also sampled upstream and downstream of the site, adjacent to the site boundaries.

3.4 Soil

During installation of the boreholes, soil samples were taken, where possible, at a depth just below first water strike. Survey was initiated in January 2003 in order to assess the extent of contamination throughout the site and to assess general groundwater movement. This survey found hydrocarbon contamination at all sample points within the site and around the periphery as shown in Table 1. A section of the site, 40 m wide and 10 m deep, is analyzed using the developed finite element model. The section is divided into 400 eight-node quadratic rectangular elements. The transport of the contaminant by diffusion and dispersion mechanisms is considered.

Fig. 2 shows, the distributions of contaminant concentration between January 2003 and September 2004. It can be seen that the contaminant concentration decreased gradually over this period of time from an initial distribution of amplitude $c = 115034\,\mathrm{mg/l}$ centered at $x = 17.5\,\mathrm{m}$. The figure also compares the results of the model prediction with the measured values of contaminant

Table 1. Analysis of contaminants in groundwater (January 2003)

Sample ID	B2	B5	B6	B7
Total Petroleum Hydrocarbons (TPH) Mg/l	115034	22000	20100	2462

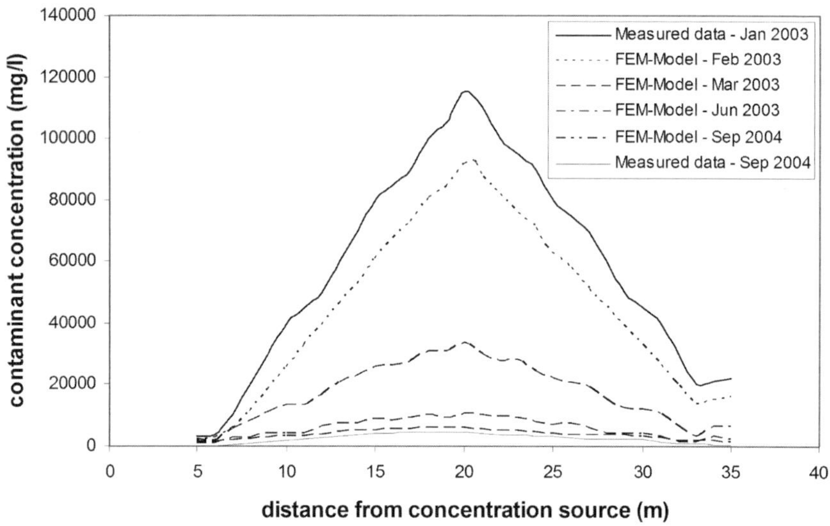

Fig. 2. Comparison between measured data and finite element model predictions

concentration recorded in September 2004. It is shown that the results of the developed model are in very good agreement with field measurements, both in terms of magnitude and trend of variations. After 16 months, the concentration of contaminant in the soil reduced by 99.3% and the slight difference between the measured and predicted concentrations could be attributed to the errors in determination of coefficients of diffusivity and permeability as well as the simplifications adopted in numerical modelling. Furthermore, considering the fact that a two-dimensional model was used to simulate a real world complex problem, such small differences in predictions are inevitable, expected and acceptable for practical applications.

4 Conclusions

This paper presented a numerical model for predicting contaminant transport through soils. The model is capable of simulating several phenomena governing miscible contaminant transport in the soils including advection, dispersion, diffusion and adsorption. A transient finite element model was developed to solve the governing equation of contaminant transport together with the equations for air and water flow. The numerical model was validated by application to a case study involving transport of a petroleum-based contaminant at a site in south west of England. The model was used to study the distribution of the contaminant with time and to evaluate the potential and degree of contamination of the site from previous commercial use as a fuel filling station and vehicle repair workshop. The results show that the developed numerical model is capable of predicting, with a very good accuracy, the effects of various mechanisms of contaminant transport through soils.

References

Hellawell EE, Sawidou C (1994) A study of contaminant transport involving density driven flow and hydrodynamic clean up. In: Centrifuge '94 Conference. University of Cambridge, U.K.

Javadi AA, AL-Najjar MM, Elkassas ASI (2006) Numerical modelling of contaminant transport in unsaturated soil. In: 5th Int Cong Env Geotechnics. Cardiff University, U.K. 1177–1184

Li X, Cescotto S, Thomas HR (1999) Finite element method for contaminant transport in unsaturated soils, J Hydrologic Engineering 4:265–274

Nazaroff WW, Alvarez-Cohen L (2001) Environmental Engineering Science. John Wiley & Sons, Inc., U.S.A.

Numerical Modelling of Flow in Variably Saturated Porous Media with Fluctuating Shallow Water Tables

Maria Mavroulidou and Michael J. Gunn

Faculty of Engineering, Science and The Built Environment, South Bank University, 103 Borough Road, London SE1 0AA, UK
mavroum@lsbu.ac.uk, gunnm@lsbu.ac.uk

Summary. The paper discusses various models which have been proposed to simulate the hysteresis of the water retention curve. Three of these models were implemented into a finite element computer program for flow in variably saturated porous media and were applied to the solution of transient flow through a soil column subject to drying-wetting cycles. The performance of these models was assessed through comparisons with previously published experimental results. A linear hysteresis model was shown to reproduce satisfactorily the experimental results, although it was simple to implement and required minimal experimental data to be formulated. Overall, the results indicate that the fact of including hysteresis in the numerical model improved the predictions with respect to the case when single-valued functions were used.

Key words: soil water characteristic curve, hysteresis, numerical modelling

1 Introduction

Fluctuations of the groundwater table result in changes in the water content and pore pressure profiles in the unsaturated soil zone. The relationship between water content and pressure profile in the unsaturated zone upon consecutive drying and wetting cycles of the soil is not unique and may be subject to considerable hysteresis. The wetting and drying behaviour of most natural soils would be expected to lie within a main drying and a main wetting curve. For several cycles of wetting and drying, scanning curves between the two main wetting and drying curves also need to be calculated.

The amount of experimental work needed for the knowledge of the rewetting re-drying scanning curves when two or more wetting-drying circles are involved is rather prohibitive. For this reason, several models for the scanning curve description have been proposed in the literature. Physically based models (the domain models), are classified into two categories, i.e. the independent domain models and the dependent domain models, based on the assumptions that they involve. According to the independent domain method (e.g. Poulovassilis 1962, Mualem 1973) the draining or filling of each pore of the domain is produced independently of the surrounding pores state and that the water volume difference between the empty and the filled state of each pore is independent of the pressure head. Only the pore geometry determines the drying and wetting characteristics of each pore. Conversely, the dependent domain models (e.g. Poulovassilis and Childs 1971, Mualem 1984) include a domain dependence factor, such that the draining and wetting of each pore be dependent on the state of the neighbouring pores (i.e. whether they are water- or air-filled). These models, introduced as an improvement to the independent domain models, are mathematically more complex and may require more measured data than the former models (Mualem 1973, Jaynes 1984)). For instance, the model by Mualem and Miller (1979) apart from the two main wetting and drying curves requires knowledge of one primary scanning curve at least. It is rare however, that such measured data are provided.

A simpler method for calculating the scanning curves is using linear-type models. Such models, proposed very early in the literature (e.g. Hanks et al. 1969), assume that the scanning curves can be approximated by straight lines, the slopes of which depend upon the main wetting and drying curves. Although crude, such models are attractive as they only require a minimum amount of input data (i.e. only the main wetting and drying curves), they are the easiest to formulate mathematically and implement in computer codes, and they are computationally less intensive.

Previous research has investigated the comparative performance of various hysteresis models based on the goodness of fit of experimentally obtained curves (e.g. Jaynes 1984, Melgarejo-Corredor 2004). This includes a very recent paper by Pham et al. (2005), which provided a thorough explanation of models for the soil water retention curve hysteresis and a comparison of five of these such models. The present paper attempts to investigate the importance of considering hydraulic property hysteresis in engineering applications involving fluctuating shallow water tables. For this, numerical results from analyses a) using a unique soil water characteristic curve and b) considering hysteresis of the soil water characteristic curve, are compared to experimental results of transient flow in a soil column. The hysteresis models used are a simple linear hysteresis model for the scanning curves as well as two more complex hysteresis models by Mualem (an independent domain model (Mualem 1973) and a dependent domain model (Mualem 1984)).

2 Description of the Numerical Model

The basic equation governing flow of water through the soil is:

$$\frac{\partial}{\partial x}\left(K_x(\psi)\frac{\partial h}{\partial x}\right)+\frac{\partial}{\partial y}\left(K_y(\psi)\frac{\partial h}{\partial y}\right)+\frac{\partial}{\partial z}\left(K_z(\psi)\frac{\partial h}{\partial z}\right)=(\beta Ss+C(\psi))\frac{\partial h}{\partial t} \tag{1}$$

where h is the total head, ψ is the pressure head and $K_x(\psi)$, $K_y(\psi)$ and $K_z(\psi)$ the hydraulic conductivities in the x, y and z directions, respectively. S_s is the elastic storage coefficient; β is equal to 1 for fully saturated conditions and 0 for unsaturated conditions and $C(\psi)$ is the moisture capacity coefficient. For the derivation of Eq. (1) flow was assumed to obey Darcy's law in both saturated and unsaturated zones.

To describe the form of the moisture retention and hydraulic conductivity curves versus pressure head, the following expressions were used:

$$\theta(\psi) = (\theta s - \theta r)\frac{A_1}{A_1 + |\psi|^{n_1}} + \theta r\,, \tag{2}$$

$$K = K_s \Theta \delta\,. \tag{3}$$

In Eq. (2) q is the volumetric water content and q_s and q_r stand for the saturated and residual volumetric content respectively; A_1 and n_1 are fitting parameters. In Eq. (3) K_s is the saturated hydraulic conductivity coefficient, d is an empirical constant estimated as 3 by Abrishami (1987) and 3.5 by Irmay (1954); Θ is the normalised volumetric water content, defined as $\Theta = (\theta - \theta_{\text{res}})/(1 - \theta_{\text{res}})$. The specific moisture capacity $C(y)$ is obtained by differentiation of Eq. (2).

Three different methods to model soil water characteristic curve hysteresis were considered and assessed, namely a linear model, an independent domain model (Mualem 1973) and a dependent domain model (Mualem 1984). Due to space limitations, a detailed description of the models and their computer implementation will not be provided. Details can be found in (Mavroulidou 1999). Note that hysteresis of the $K(\theta)$ curve was not considered, as this is usually shown to be negligible (Topp 1971).

3 Numerical Analyses

3.1 Geometry, Material Properties and Boundary Conditions

To check the importance of accounting for hysteresis in engineering applications, experimental results found in Abrishami (1987) were checked against numerical results. These results were for transient, one-dimensional flow in a 34 cm medium sand column with variable boundary conditions (causing alternating drying and wetting of the porous medium). The material properties were: $q_s = 0.359$, $q_r = 0.02$ and $K_s = 0.106\,\text{cm/s}$. In Eq. (2), $A_1 = 12 \times 10^6$ and $n_1 = 5.82$ for the drying curve, as fitted by Abrishami (1987). For the linear model, scanning curves obtained experimentally by Abrishami (1987) were used to find the slope of the scanning lines. The slope of the intermediate scanning curves was assumed to be the same as that of the nearest overlying experimentally measured scanning curve.

The initial and boundary conditions were the following:

$t = 0,$	$0\,\mathrm{cm} \le z \le 34\,\mathrm{cm},$	$h = 17.6\,\mathrm{cm};$
$t \ge 0,$	$z = 34\,\mathrm{cm},$	$\dfrac{\partial h}{\partial z} = 0$

<u>First drying:</u>

$t = 1\,\mathrm{min},$	$z = 0\,\mathrm{cm},$	$h = 11.8\,\mathrm{cm};$
$t = 2\,\mathrm{min},$	$z = 0\,\mathrm{cm},$	$h = 5\,\mathrm{cm};$
$2.5\,\mathrm{min} \le t \le 14\,\mathrm{min},$	$z = 0\,\mathrm{cm},$	$h = 0\,\mathrm{cm};$

<u>First re-wetting:</u>

$t = 15\,\mathrm{min},$	$z = 0\,\mathrm{cm},$	$h = 12\,\mathrm{cm};$
$15.5\,\mathrm{min} \le t \le 24\,\mathrm{min},$	$z = 0\,\mathrm{cm},$	$h = 17.6\,\mathrm{cm};$

<u>First re-drying:</u>

$t = 25\,\mathrm{min},$	$z = 0\,\mathrm{cm},$	$h = 8.5\,\mathrm{cm};$
$t \ge 26\,\mathrm{min},$	$z = 0\,\mathrm{cm},$	$h = 0\,\mathrm{cm}$

3.2 Presentation and Discussion of Results

Due to space limitations only indicative numerical results will be presented. A more detailed description can be found in Mavroulidou (1999). The numerical results for the first drainage (using the drying curve) show satisfactory agreement with the experimental results (Fig. 1).

However, for subsequent re-wetting, the results based on a unique curve (no hysteresis) show faster rates of evolution than the experimental results. Conversely, when hysteresis is considered the results are in closer agreement with the experimental results (Fig. 2), with the linear model being slightly better. Note that only one of Mualem's models is represented (Mualem 1973) in Fig. 2 as in this instance the differences between the results derived by the two Mualem's models were very small.

Similar observations can be made for the first re-drying results: when the drying curve is used throughout the analysis the results tend to show faster rates of evolution than those found form the experimental results. The linear model gives in general results closer to the experimental. In this case, the performance of Mualem's models was not as good as in the first re-wetting case. However, the hysteresis models still gave overall results closer to experimental observations than the those derived from the drying curve (i.e. when no hysteresis was accounted for). Again only one of Mualem's models is shown in Fig. 3 as overall, the differences were very small, except at the last two time levels. Note that when Mualem's (1984) method was used, some slight numerical instabilities in the results were observed for the last time level.

Based on the results, the overall behaviour of the linear model for the scanning curves is judged to be satisfactory. Conversely, the accuracy of Mualem's models for the secondary scanning curves needs further investigation. The better relative performance of the linear model may be surprising as the

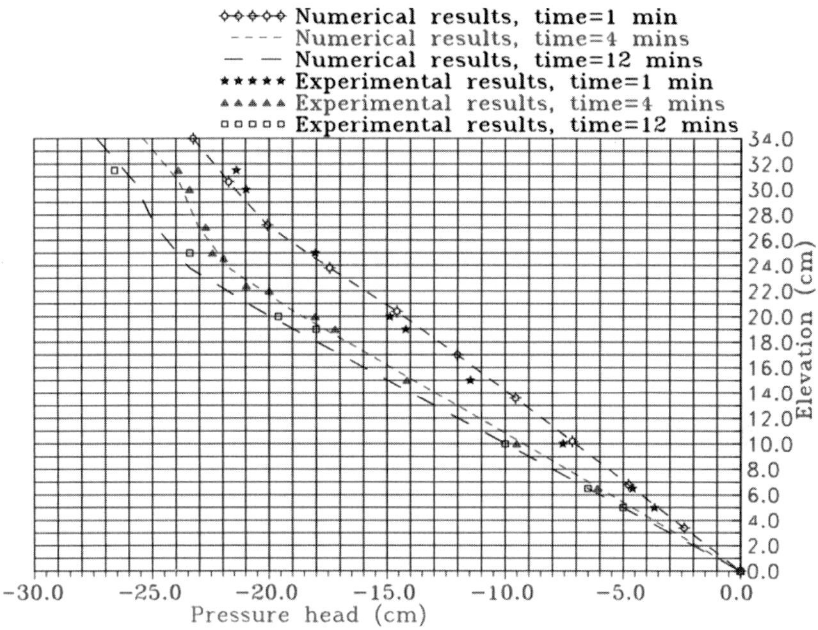

Fig. 1. Transient flow through 1-D sand column: Initial drainage

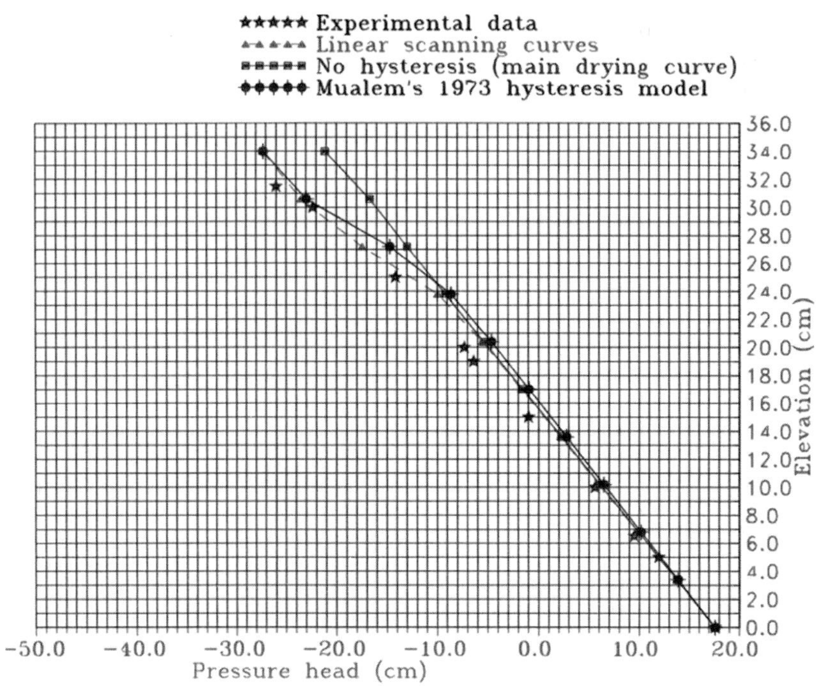

Fig. 2. Transient flow through 1-D sand column: First re-wetting, $t = 16\,\text{min}$

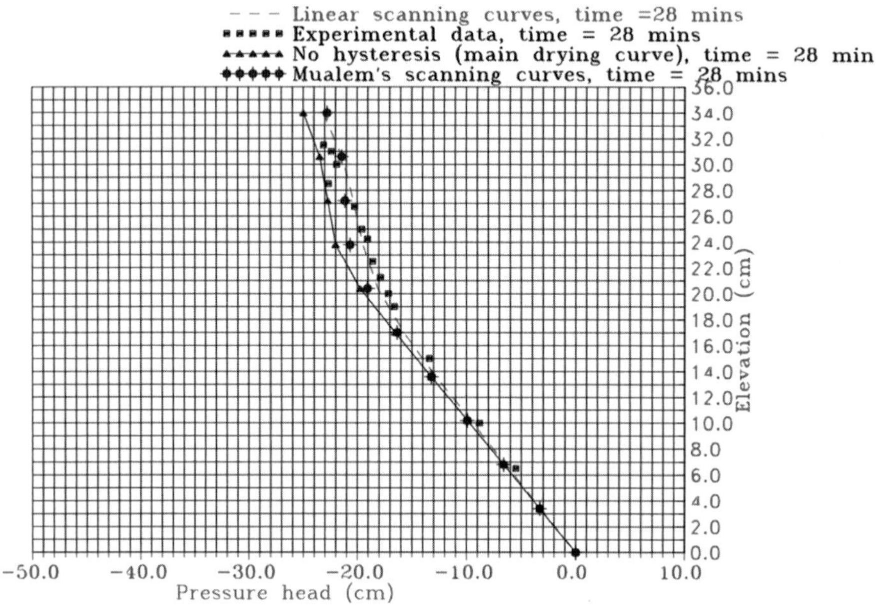

Fig. 3. Transient flow through 1-D sand column: First re-drying, $t = 28\,\mathrm{min}$

representation of the scanning curves by straight lines is crude. The results however confirm previous observations Jaynes (1984), that more sophisticated hysteresis models (e.g. Mualem 1974) did not provide a good representation of the scanning lines. Additionally, a "pumping effect" was observed for some more complex models (e.g. Dane and Wierenga 1975) as opposed to the linear model, i.e. a decrease in the predicted soil-water contents during the initial pressure fluctuations until some stable water content configuration is reached. This is a numerical artefact rather than a real property of the soils

It should be noted that the results and related conclusions refer to shallow water tables. Moreover, the porous medium was assumed to be rigid.

4 Conclusions

The results point at the importance of including hysteretic behaviour of the medium in models of flow through variably saturated soils. The linear hysteresis model was shown to reproduce satisfactorily experimental results of seepage, and perform better than more sophisticated models. The use of the linear model in numerical models could therefore be advantageous, as it is simple, computationally less intensive than other models and does not require a large amount of experimental data to be formulated.

References

Abrishami J (1987) An Investigation into the Flow of Water in Partially Saturated Soils and its Representation Using Finite Elements. Ph.D. Thesis, University of Wales Cardiff

Averjanov SF (1950) About permeability of subsurface soils in case of incomplete saturation. Eng. Collect. 7

Dane JH, Wierenga PJ (1975) Effect of hysteresis on the prediction of infiltration, redistribution and drainage of water in a layered soil. J. of Hydrol 25:229–242

Hanks RJ, Klute A, Bresler A (1969) A numeric method for estimating infiltration, redistribution, drainage and evaporation of water from soil. Water Res. Res. 5:1064–1069

Irmay S (1954) On the hydraulic conductivity of unsaturated soils. Transactions of the American Geophysical Union 35:463–468

Jaynes DB (1984) Comparison of soil-water hysteresis models. J Hydrol 75:287–299

Mavroulidou M (1999) Numerical modelling of seepage in the presence of phreatic surfaces. Ph.D. Thesis, University of Surrey Guildford

Melgarejo-Corredor ML (2004) Laboratory and numerical investigation of soil retention curves. Ph.D. Thesis, Imperial College, London

Mualem Y (1973) Modified approach to capillary hysteresis based on similarity hypothesis. Water Res. Res 9:1324–1331

Mualem Y (1974) A conceptual model for hysteresis. Water Res. Res 10:514–520

Mualem Y (1984) A modified dependent-domain theory of hysteresis. Soil Sci 137:283–291

Mualem Y, Miller EE (1979) A hysteresis model based on an explicit domain dependence function. Soil Sci Soc Amer 43:1067–1073.

Pham HG, Fredlund DG, Barbour SL (2005) A study of hysteresis models for soil-water characteristic curves. Can. Geotech. J. 42:1548–1568

Poulovassilis A (1962) Hysteresis of pore-water, an application of the concept of independent domains. Soil Sci 93:405–412

Poulovassilis A, Childs EE (1971) The hysteresis of pore-water: the nonindependence of domain. Soil Sci 112:301–312

Topp GC (1971) Soil water hysteresis in silt loam and clay loam soils. Water Res. Res 7:915–920

Residual Strength of Clays at High Suctions

Jean Vaunat, Vladimir Merchán, Enrique Romero, and Jubert Pineda

Department of Geotechnical Engineering and Geosciences,
Technical University of Catalonia (UPC), Building D2, Campus Nord,
c/ J. Girona, 1-3, 08034 Barcelona, Spain
jean.vaunat@upc.edu, vladimir.merchan@upc.edu,
enrique.romero-morales@upc.edu, jubert.pineda@upc.edu

Summary. This paper presents a study on the residual shear strength of a plastic clay (Boom clay) under high suction. The device, an adaptation of Bromhead ring shear apparatus to allow for suction control during shearing, is first described and the experimental procedure detailed. Then, results of shear strength measured under a suction equal to 70 MPa, several vertical stresses and different shearing history (first-shearing and shearing on pre-sheared samples) are presented and discussed. They evidence a huge increase in the residual friction angle whose value is 15° greater than that measured in saturated conditions and a highly dilatant response during first-shearing. ESEM micrographs performed on samples sheared in saturated and unsaturated conditions suggest that such a response can be explained by the character more granular of the dry material as a result of particle aggregation. Comparison with results obtained by Vaunat et al. (1981) on a low-plastic clay indicate that this effect is enhanced by the plasticity of the clay.

Key words: unsaturated soil, residual strength, laboratory testing

Introduction

The importance of residual strength of soils is well-recognized in problems like reactivation of landslides (Skempton 1964), pile friction resistance (Lehane and Jardine 1992), foundations on stiff clays, earth dams (Alonso 1998), etc. Most of the experimental work done in this domain refers to the testing of clayey materials in the direct shear box, the conventional (Bromhead 1979) and modified (Anayi et al. 1989) Bromhead ring shear apparatus. It is now of general acceptance that the residual strength in saturated conditions is characterized by null cohesion and a friction angle that depends on the level of normal stress (Skempton 1985, Stark and Eid 1994), soil grading (Skempton 1964, Kenney 1967, Lupini et al. 1981, Skempton 1985), particle mineralogy (Mitchell 1993), rate of shearing (Tika et al. 1996) and pore water chemistry (Di Maio 1996a,b, Chighini et al. 2005). Residual strength of unsaturated clay has been surprisingly never studied, maybe because of the belief that the mechanical effect of suction is due to the existence of water menisci and that such menisci do not exist across shear surface characterized by large relative displacements. Notwithstanding, Vaunat et al. (1981) have shown in a recent work that the friction angle of a low plastic clay may experiment an increase of 5° when the soil is subjected to high suctions.

This paper is devoted to a continuation of this study. It presents and discuss experimental results obtained on a medium plastic clay subjected to high suctions. The device is essentially the same as the one proposed by Vaunat et al. (1981) but with a new system to isolate the box from the room atmosphere. Material tested is Boom clay, a tertiary clay with LL = 55% and $I_p = 28\%$.

In a first part, the paper presents the features of the suction-controlled ring shear apparatus. Afterwards the preparation of the material and the procedure followed to test the clay are detailed. Results obtained for the residual strength of the clay are then described and discussed in the light of the change in microstructure that occurs when the soil is dried. Some preliminary conclusions are finally drawn.

Test Procedures

The test device is a Bromhead apparatus adapted to control suction by vapour transfer during the shearing stage (Vaunat et al. 1981). Figure 1 shows a picture of the general system. The shear box is isolated from the ambient atmosphere by a glass-cap and connected to a circuit of forced convection of vapour. Relative humidity (Rh) is controlled by circulating the air above a solution saturated in salts of different types. Relativity humidity and temperature in-

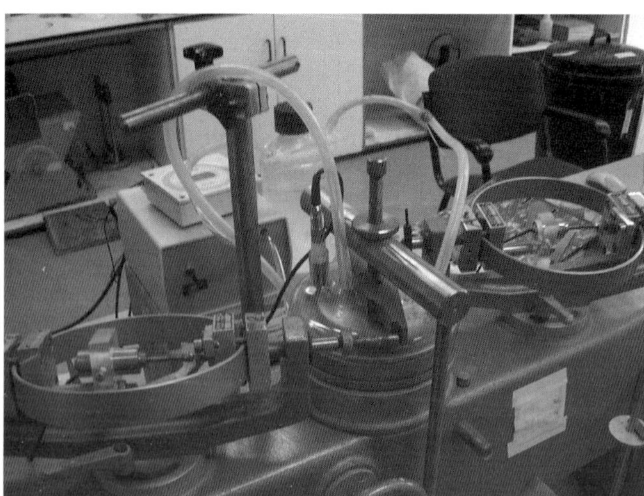

Fig. 1. Adapted general system to control suction on Bromhead ring shear apparatus

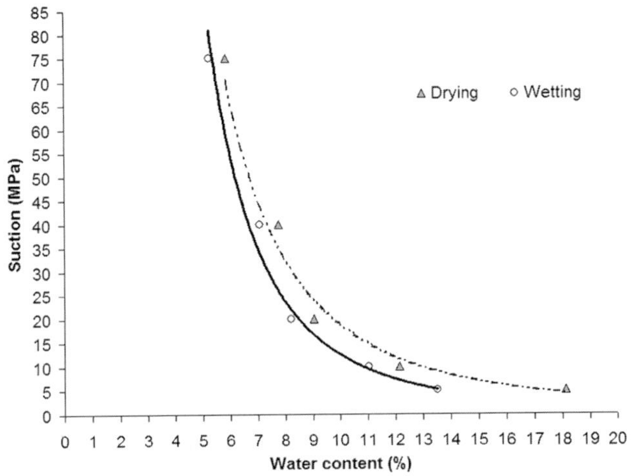

Fig. 2. Retention curve of Boom clay (see Lima et al. 2006)

side the chamber are registered by a transmitter HMT330 (accuracy of ±1% between 0 and 99% and ±2% between 99 and 100%) during the entire test together with the vertical displacement of the upper box. Shear resistance is measured during the shearing stage. The material tested is a clay of medium plasticity (Boom clay). Its liquid limit is equal to 55%, plasticity index 28% (Lima et al. 2006) and particle density 2.7 Mg/m^3. The retention curve of the material for drying and wetting paths (Lima et al. 2006) is shown in Fig. 2. It is worth noting that suction changes for values higher than 35 MPa are related to little variations in gravimetric water content.

Sample is prepared at its plastic limit in the remoulded state. It is placed in the shear box and left equilibrates under a given vertical stress. Afterwards suction is imposed by putting on the vapour convection system.

Equilibrium Stage

Figure 3 shows and example of stabilization evolution for a sample loaded at 37.5 KPa of normal stress. Three different stages can be recognized.

The loading stage characterized by a short consolidation stage of one day.

The application of suction, done by connecting the air circuit to a recipient with a relative humidity equal to 75% (value prevailing above saturated aqueous solution of NaCl). During this stage, sample experiments an important vertical compressive strain equal to 14%. It is assumed that suction equilibration is achieved when compression stabilizes, that is almost 14 days after suction application.

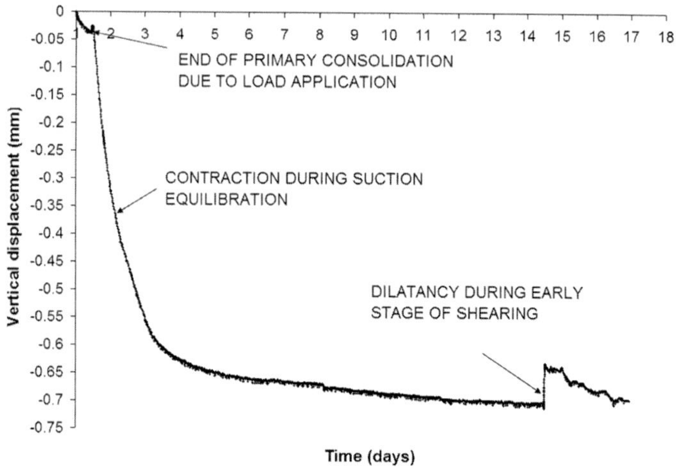

Fig. 3. Vertical displacement evolution during consolidation, suction equilibration and early stage of shearing

The shearing stage, performed at a rotational speed of 0.12°/min during four days. Sample exhibit a sudden high dilatancy during the early stage of shear process.

Shearing Stage

During shearing, the relative humidity and the temperature were measured inside the glass-cap. Results indicate that the average relative humidity inside the shear box is typically 20% lower than that prevailing above the NaCl solution and fluctuates with time. The variation of Rh measured during shearing on a pre-sheared sample under a vertical stress equal to 37.5 kPa is shown in Fig. 4. Values oscillate between 47% to 59%. Such a fluctuation is accompanied by a variation in suction that reaches temporally 30 MPa. Nevertheless, the value of shear strength measured during the same interval of time has remained very stable. This surprisingly low variation in strength can attributed to two reasons: 1) according to the retention curve, the change in water content for suction varying between 60 and 90 MPa is very small (ω remains between 5 and 6%); 2) Rh fluctuations are too fast to develop properly inside the sample. The value of the shear strength is thus considered to be essentially controlled by the suction prevailing at the end of the equilibration stage that is 70 MPa in average.

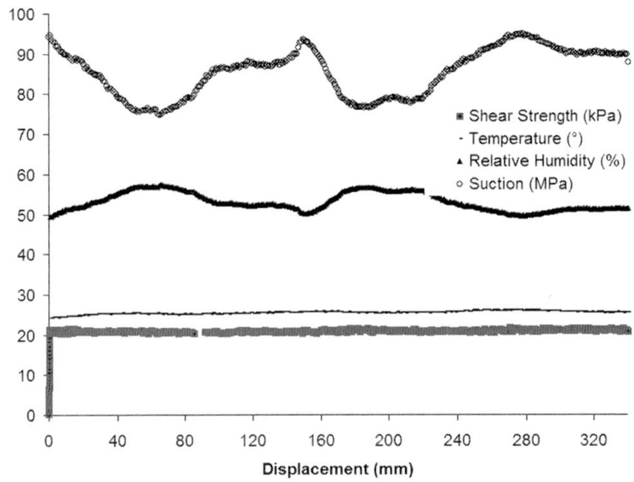

Fig. 4. Relative humidity, temperature, suction and shear strength evolution for pre-sheared sample at 37.5 kPa of normal stress

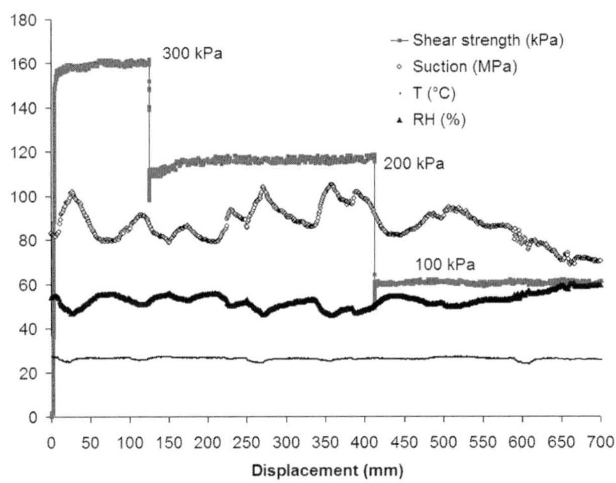

Fig. 5. Relative humidity, temperature, suction and shear strength evolution for pre-sheared sample at multi-load steps

Test Results

Two tests were realized at an average constant suction equal to 70 MPa. During the first test (Test A), seven shearing stages (numbered from A1 to A8 in the remaining text) were conducted sequentially under vertical stresses equal to 300, 200, 100, 37.5, 87.5, 100, 200 and 300 kPa, respectively. During the second test (Test B), sample was sequentially sheared by stages under a vertical stress equal to 87.5 (stage B1), 37.5 (stage B2) and 87.5 kPa (stage B3). Total displacement applied during each stage is shown in Fig. 6.

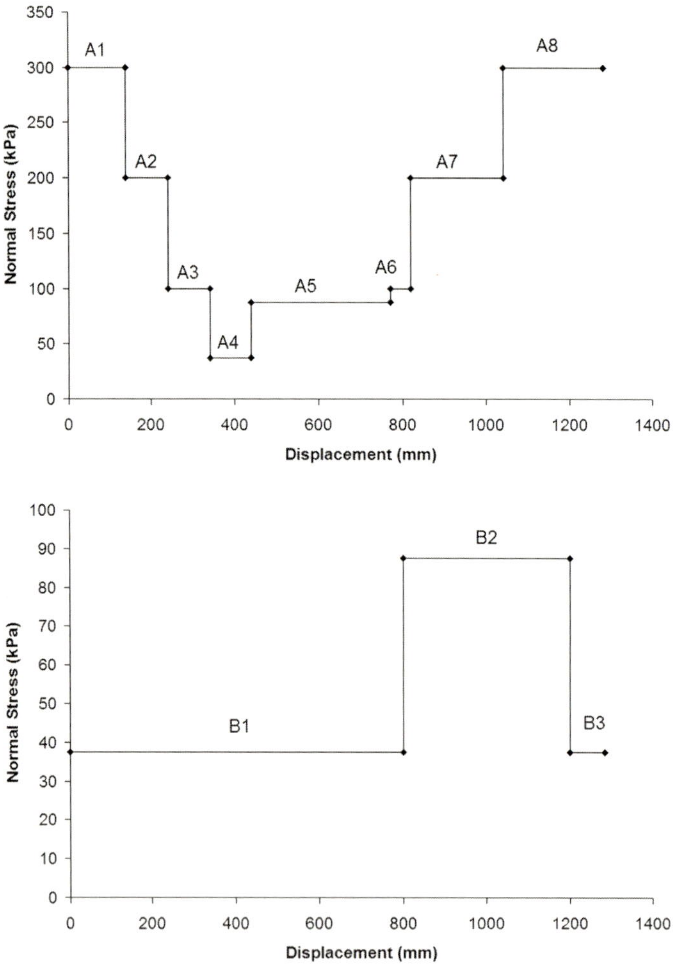

Fig. 6. Displacement-vertical load path applied during the tests

Results of the tests are depicted in Figs. 7, 8 and 9. During first-shearing, both samples exhibit a dilatancy and brittle behaviour (Stage A1 in Fig. 7, Stage B1 in Fig. 8). The loss of strength measured during Test A under a vertical stress $\sigma'_n = 300\,\text{kPa}$ is equal to 240 kPa. Corresponding dilatancy appears to be very high: a vertical upward displacement equal to 0.25 mm, which is 5% of the height of the sample is registered exactly at time of peak development.

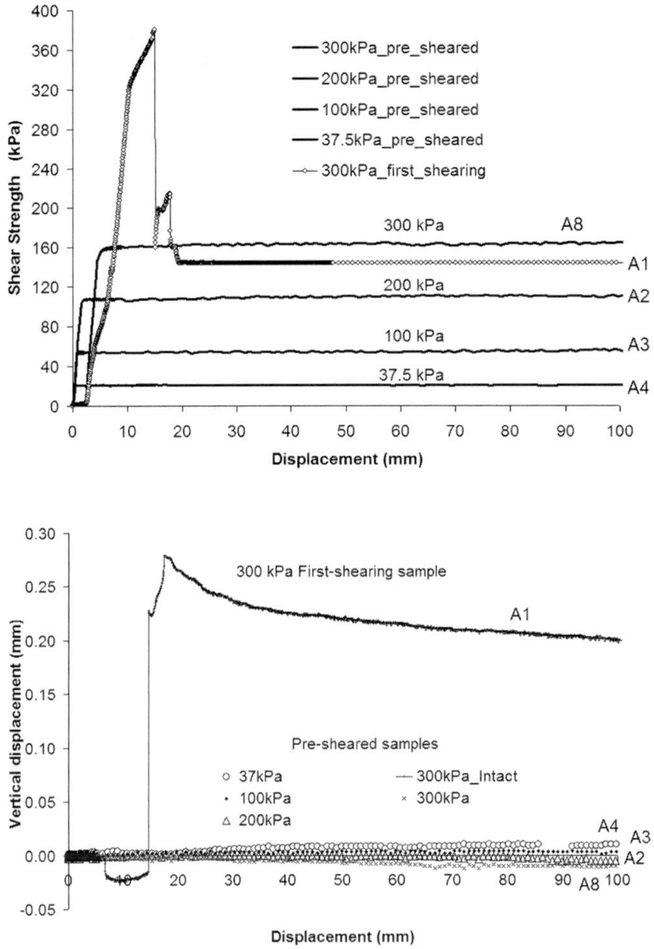

Fig. 7. Test A: shear stress and vertical displacement vs horizontal displacement for the first shearing stage under $\sigma_n = 300$ and the following stages under $\sigma'_n = 200$, 100, 37.5 and 300 kPa

It has been impossible to capture the softening branch in this test, because both the drop in strength and the dilatancy occurred abruptly between two consecutive measurements, separated only by a time interval of 1 s (increment of displacement 0.05 mm). Afterwards, strength stabilizes while displacement slowly decreases of some percent of millimetre. A similar response is observed during the first-shearing stage of Test B ($\sigma'_n = 37.5$ kPa). As shown in Fig. 8, the loss of strength is however more progressive: a decrease of 12 kPa is registered during a displacement increment equal to 2.7 mm. These observations are consistent with those realized on Barcelona silty clay ($I_p = 14\%$) at different

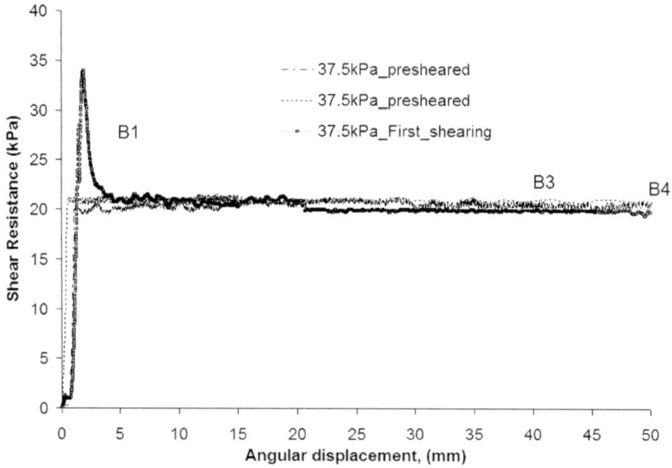

Fig. 8. Shear stress vs displacement curve for $\sigma'_n = 37.5\,\text{kPa}$

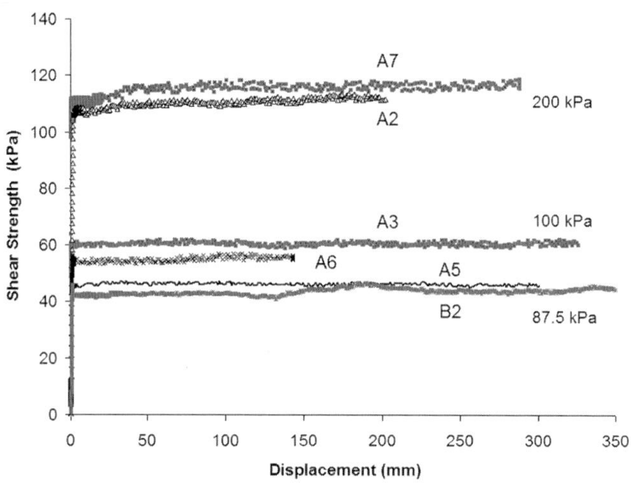

Fig. 9. Shear stress vs displacement curve for $\sigma'_n = 87.5$, 100 and 200 kPa

suctions (Vaunat et al. 1981). Tests on both materials evidence the enhancement of dilatancy and brittleness by suction increase during first-shearing.

When shearing is applied on a pre-sheared sample (Stage A2, A3, A4 and A8 in Fig. 7, Stage B2 and B3 in Fig. 8), the dilatant and peak response of the material disappears. Comparison with results of first-shearing indicates moreover that the value of the residual strength measured on pre-sheared sample is close to that obtained at the end of first-shearing under the same vertical stress (see comparison between stage A1 and A8 in Fig. 7, Stage B1 and B3 in Fig. 8). This fact, already quoted in Vaunat et al. (1981) for the

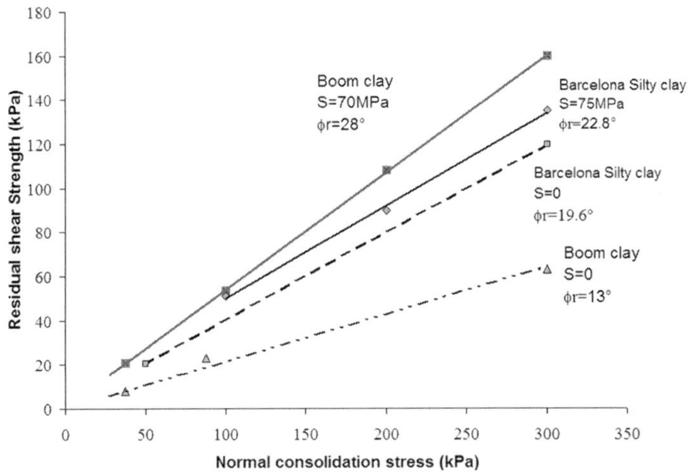

Fig. 10. Residual shear resistance envelopes for saturated and unsaturated Boom clay and Barcelona silty clay

case of a Barcelona silty clay, indicates that the residual strength at a given suction depends only on the applied normal stress and not on the shearing history. The comparison of the shear strength between two stages performed on pre-sheared sample with the same vertical stress but different displacement history is shown in Fig. 9. It confirms the independence of the residual strength on the shearing history for the samples studied.

The residual shear envelopes of Boom clay at a suction of 70 MPa is plotted in Fig. 10. On the same figure, the residual strength measured in the same apparatus on a saturated sample is indicated. Results are compared with those obtained by Vaunat et al. (1981) on a Barcelona silty clay (CF = 30%, I_p = 14%). Both envelopes are characterized by a null cohesion and a friction angle ϕ'_r that appears to be constant for the range of normal stress applied during the tests. An appealing result is the huge increase of the residual friction angle experimented by the clay during drying. From this first results, ϕ'_r appears to change from 13° in saturated conditions to 28° at 70 MPa of suction. On the other hand, Barcelona silty clay plasticity clay appears to have an increment from 19.6° in saturated conditions to 22.8° at 75 MPa of suction. Both results suggest that suction plays an important role for the residual strength of clay and that this effect is highly conditioned by the plasticity of the clay.

Two ESEM micrographs of Boom clay material are shown in Fig. 11. The scale indicated in each picture is 25 nm, which is of the order of clay aggregates (Chighini et al. 2005). The saturated material sheared in saturated conditions (Fig. 11**a**) presents a relatively homogeneous structure with no clear development of inter-aggregate porosity. Sample sheared under 70 MPa of suction presents a different picture. Large interstices isolating clusters appear to be developing, in a process that is expected to come out into the generation of clay aggregation. Such a change in microstructure would have an effect on both the grading and the plasticity of the clay.

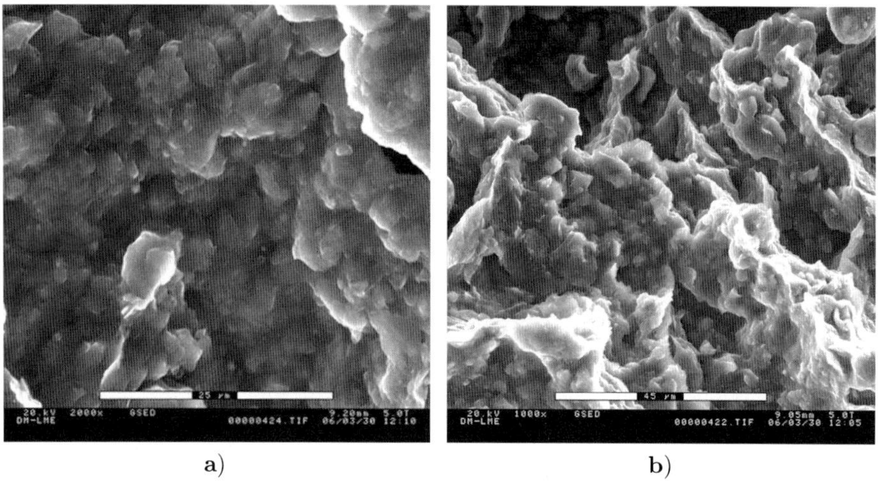

Fig. 11. ESEM micrographs of **a**) saturated and **b**) unsaturated Boom clay after shearing

The residual friction angles measured on saturated and unsaturated samples of Boom-clay and Barcelona silty clay are indicated on Fig. 12 reported by Lupini et al. (1981), that show the correlation obtained by several authors between ϕ_r', the clay fraction and the plasticity index of a wide range of clay materials. Points corresponding to the saturated sample of Boom clay and Barcelona silt in Fig. 12**a** are close to the lower bound proposed by Skempton (1985) while points corresponding to unsaturated samples are located near the upper bound. This seems to indicate that the material becomes more granular during the drying process, which would also give an explanation to the dilatant response observed on unsaturated sample during first-shearing. A similar conclusion can be drawn from the Fig. 12**b**. In this case, application of suction appears to have effect similar to that of a reduction in the plasticity of the material.

Conclusions

The Bromhead ring shear apparatus have been adapted to control suction during shearing and a series of tests realized on a medium plastic clay (Boom clay) in saturated and unsaturated conditions. Results indicate that:

- Residual strength in unsaturated conditions is independent of the stress history.

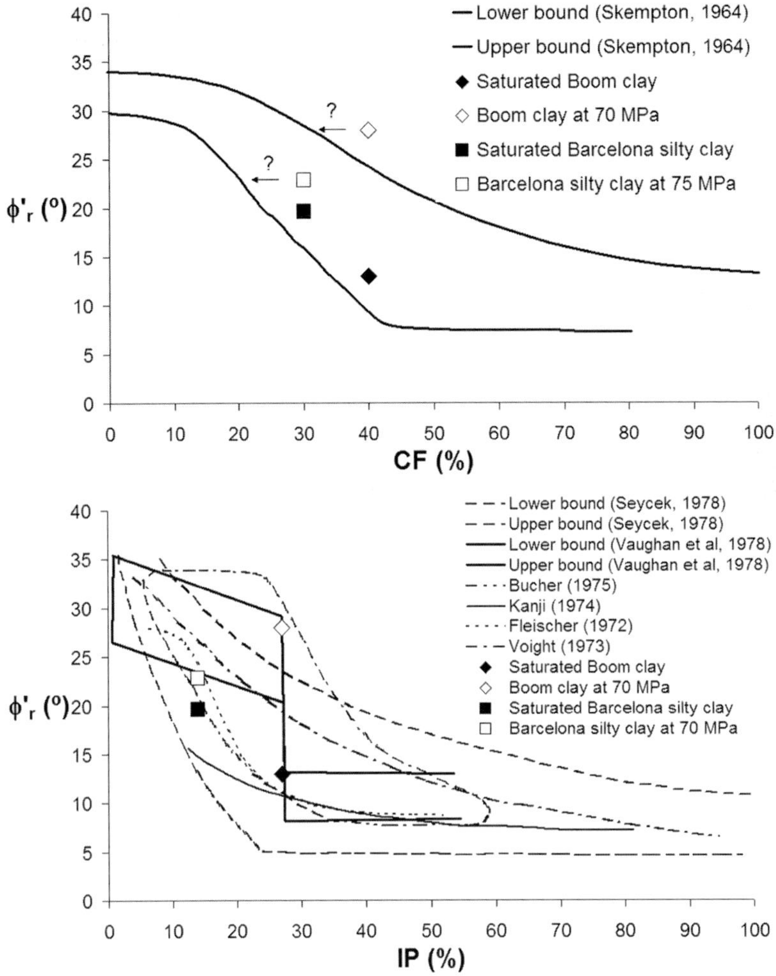

(from Lupini et al. (1981))

Fig. 12. Plot on the figure reported by Di Maio (1996b) of the relationship between residual friction angle and **a**) clay fraction, **b**) plasticity index for the samples of Boom clay and Barcelona silty clay in saturated and unsaturated conditions

- During first shearing, sample is contractant in saturated conditions and highly dilatant for a suction equal to 70 MPa. Dilatancy disappears when the sample is pre-sheared.
- The increase of the friction angle due to suction is huge (+15°) and much higher than that observed on a low plasticity clay (Barcelona silty clay) by Vaunat et al. (1981).

These results suggest that suction application develops aggregation of particles, which leads the sample to behave like a more frictional material during shearing. This hypothesis could explain the increase in residual friction angle observed during the unsaturated tests as well as the occurrence of high dilatancy during first-shearing. Further research is necessary to analyze in detail the change in clay grading and plasticity due to drying and its consequence in terms of mobilized strength.

Acknowledgements

Mr. Merchán is supported by the Programme Alβan, the EU Programme of high level Scholarship for Latin America, scholarship No E05D052296CO.

References

Alonso EE (1998) General report. Cutting and natural slopes. In: Evangelista, Picarelli (eds) Proc 2nd Int Symp *The geotechnics of hard soils – soft rocks.* Balkema, Rotterdam:1441–1481

Anayi JT, Boyce JR, Rogers CDF (1989) Modified Bromhead ring shear apparatus, Geotech Test J 12(2):171–173

Bromhead EN (1979) A simple ring shear apparatus, Ground Engineering 12(5):40–44

Chighini S, Lancellotta R, Musso G, Romero E (2005) Mechanical behavior of Monastero Bormida clay: chemical and destructuration effects. In: Bilsel, Nalbantoðlu (eds) Proc Int Conf on Problematic Soils. Eastern Mediterranean University, Famagusta 1:381–388

Collins K, McGown A (1974) The form and function of microfabric features in a variety of natural soils, Géotechnique 24(2):223–254

Di Maio C (1996a) The influence of pore fluid composition on the residual shear strength of some natural clayey soils. In: Senneset K (ed) Proc 7th Int Conf on Landslides. Balkema, Rotterdam 2:1189–1194

Di Maio C (1996b) Exposure of bentonite to salt solution: osmotic and mechanical effect, Géotechnique 46(4):695–707

Kenney TC (1967) The influence of mineral composition on the residual strength of natural soils. In: Proc Geotech Conf on the Shear strength properties of natural soils and Rocks 1:123–129

Lehane BM, Jardine RJ (1992) Residual Strength characteristics of Bothkennar clay, Géotechnique 42(2):363–367

Lima A, Romero E, Gens A (2006) Annual report: Thermo-hydro-mechanical behaviour of saturated and unsaturated natural Boom clay

Lupini JF, Skinner AE, Vaughan PR (1981) The drained residual strength of cohesive soils, Géotechnique 31(2):181–213

Mitchell JK (1993) Fundamentals of soil behaviour, 2nd ed. John Wiley & Sons, New York

Skempton AW (1964) Long-term stability of clay slopes, Géotechnique 14:77–101

Skempton AW (1985) Residual strength of clays in landslides, folded strata and the laboratory, Géotechnique 35(1):3–18

Stark TD, Eid HT (1994) Drained residual strength of cohesive soils, J Geotech Engng, ASCE 120(5):856–871

Tika TE, Vaughan P, Lemos LJLJ (1996) Fast shearing of pre-existing shear zones in soil, Géotechnique 46(2):197–233

Vaunat J, Amador C, Romero E, Djéran Maigre D (1981) Residual strength of low plasticity clay at high sutions. In: Proc 4th Int Conf on Unsaturated Soils 1:1279–1289

Parallel Finite Element Analysis of THM Coupled Processes in Unsaturated Porous Media

Wenqing Wang[1], Panos Adamidis[2], Matthias Hess[1], Dany Kemmler[1], and Olaf Kolditz[1]

[1] GeoSystemsResearch, Center for Applied Geoscience, University of Tübingen, D-72072, Germany wenqing.wang@uni-tuebingen.de, kolditz@uni-tuebingen.de

[2] HLRS – Höchstleistungsrechenzentrum Stuttgart, D-70550 Stuttgart, Germany

Summary. Numerical analysis of thermo-hydro-mechanical (THM) coupled problems leads to an extremely high computational expense. This work is devoted to reduce such computational expense by parallel computing methods. To this purpose, parallelization is applied to the most time consuming portions of finite element simulations, i.e. assembly of linear equation systems (LES) and solving them. Since an iterative solver is adopted in the present study, the sub-structuring technique of domain decomposition plays an important role in both assembly and solving of LES. Three principles of the parallelization procedure are: (1) all processes of a coupled problem share a unique finite element mesh; (2) this mesh is discretized into sub-domains, each of them are established with mesh topology for both linear and quadratic interpolation, the assembly of linear equation systems is performed in sub-domains and is distributed to the involved processors (CPU-nodes); (3) matrix-vector multiplications, which are the basic computational operations in an iterative solver, are split to sub-domain level and are also performed by the involved processors concurrently. The parallel FEM is applied successfully to the solution of a THM coupled problem in partially saturated bentonite which are used as buffer material in geotechnical sealings.

Key words: parallel computing, finite element method, thermo-hydro-mechanical (THM) coupled problems

1 Introduction

Thermo-hydro-mechanical (THM) coupled processes in porous media are very important in many geotechnical engineering problems (Schanz 2004). A representative example is nuclear waste disposal. Nuclear waste repositories are constructed in deep geologic underground. Normally, the radioactive waste will generate heat for a long period of time with temperatures over 100°C.

Possibly, subsurface fluid flow may be developed and gas will be produced due to the heating effects. The coupling of thermal and hydraulic processes can cause mechanical damage in the near field of the host rock mass. To assess the safety of the underground repositories, the problem needs to be addressed as a thermo-hydro-mechanical (THM) coupled problem (Lewis and Schrefler 1998, de Boer 2005, Stephanson et al. 2003, Ehlers and Bluhm 2002, Kolditz et al. 2004). Although some commercial tools are already available, there is a tremendous demand in the development of fully coupled and, in particular, efficient THM codes which are able to deal with real world problems.

From the mathematics point of view, the coupled processes lead to a coupled initial-boundary-value-problems (IBVP) and the numerical simulation comes down to solve such multi-field IBVP approximately. Among the available numerical methods, the finite element method enjoys a firm theoretical foundation that is mostly free of ad hoc schemes and heuristic numerical approximations and therefore is so far the most employed numerical method to treat THM coupled problems. However, the computational costs must be taken into consideration when conducting simulations of THM coupled problems because of the increased degree of freedom and because of the strongly nonlinear behavior. There are several ways to improve the computational efficiency such as developing novel and efficient computing techniques, optimizing memory management in the code, and parallelization techniques. Among them, parallel computing provides the most powerful speedup for finite element simulations (Topping and Khan 1996). Thanks to the decreasing hardware cost in the past years, the parallel computation becomes much more attractive for research in computer science (Salinger et al. 1994, Fujisawa et al. 2003, Shioya and Yagawa 2005, Tezduyar and Sameh 2006).

In this work we present a parallel finite element method in order to improve the computational efficiency of numerical THM simulations.

2 Mathematical Background of THM Coupled Problems

Thermo-hydro-mechanical (THM) coupled processes in partially saturated porous media are described by the following initial boundary value problems (IBVP).

2.1 Flow process

We consider a general case of a flow problem in unsaturated deformable porous media. With the classical Darcy's law, the balance equations of fluid phase mass are given by

$$\frac{\partial(nS^\gamma\rho^\gamma)}{\partial t} - \nabla\cdot\left[nS^\gamma\left(\frac{k^\gamma_{rel}\mathbf{k}}{\mu^\gamma}(\nabla p^\gamma - \rho^\gamma\mathbf{g})\right)\right] + \nabla\cdot(nS^\gamma\rho^\gamma\dot{\mathbf{u}}) = Q^\gamma_f \quad (1)$$

for any point $\boldsymbol{x} \in \Omega \in \mathbb{R}^{\mathrm{n}}$ with n dimension of the real space, γ gaseous or liquid phase, S^γ saturation, p pressure, ρ density, n effective porosity of porous media, μ fluid viscosity, $\mathbf{k}$ intrinsic permeability tensor, k^γ_{rel} relative permeability, $\mathbf{g}$ gravity acceleration and Q^γ_f denotes source/sink terms. The unknown field functions of eqn. (1) to be solved are fluid phase saturation S^γ and fluid phase pressure p^γ. Fluid mass balance is coupled to the deformation of the porous medium (third term in eqn. (1)). $\mathbf{u}$ is solid displacement.

With constitutive equations for fluid density, capillary pressure and relative permeability, the balance equations (1) are closed (Kolditz 2002). The boundary conditions for this problem can be Neumman type as

$$\mathbf{q}^\gamma_f \cdot \mathbf{n} = q^\gamma_\Gamma\,, \quad \forall\, \boldsymbol{x} \in \partial\Omega \tag{2}$$

or Dirichlet type as

$$p^\gamma = p^\gamma_\Gamma\,, \quad S^\gamma = S^\gamma_\Gamma\,, \quad \forall\, \boldsymbol{x} \in \partial\Omega\,. \tag{3}$$

This type of IBVP can be solved with the corresponding initial conditions of unknowns.

2.2 Heat Transport Process

Consider advective and diffusive fluxes, the heat transport in multi-phase porous media is given by

$$\sum_\gamma^{phase} (\rho^\gamma C^\gamma_p)\frac{\partial T}{\partial t} - \nabla\left[K_e \nabla T + n\sum_\gamma^{phase}(\rho^\gamma C^\gamma_p)T\mathbf{v}\right] = Q_{\mathrm{T}} \tag{4}$$

where K_e is the heat conductivity, T is temperature, C^γ_p is specific heat capacity of fluid phase, $\mathbf{v}^\gamma$ is fluid phase velocity and Q_{T} are heat sources. The boundary conditions are given by

$$\mathbf{q}_{\mathrm{T}} \cdot \mathbf{n} = \mathbf{q}_{\mathrm{T}}|_\Gamma\,, \quad \text{or} \quad T = T_\Gamma\,, \quad \forall\, \boldsymbol{x} \in \partial\Omega\,. \tag{5}$$

The initial conditions is defined by

$$T(\boldsymbol{x}) = T_0(\boldsymbol{x})\,, \quad \forall\, \boldsymbol{x} \in \Omega\,. \tag{6}$$

2.3 Deformation Process

Deformations in porous media can be described by the momentum balance equation in the terms of stress as (Lewis and Schrefler 1982)

$$\nabla \cdot \left(\boldsymbol{\sigma} - \sum_\gamma^{phase} S^\gamma p^\gamma \mathbf{I}\right) + \rho\mathbf{g} = 0 \tag{7}$$

where $\boldsymbol{\sigma}$ is the effective stress of porous medium, $\mathbf{I}$ is identity tensor. In the present study, the traditional sign convention for stress and fluid pressure is used. Density of porous media consists of the portion contributed by liquid l and by the portion contributed of solid as $\rho = n\sum_{\gamma}^{phase} \rho^\gamma + (1-n)\rho^s$. Displacement $\mathbf{u}$ is the primary variable to be solved by substituting the constitutive law for stress-strain behavior

$$\begin{aligned} \boldsymbol{\sigma} &= \mathbb{C}\,\boldsymbol{\epsilon} \\ \boldsymbol{\epsilon} &= \frac{1}{2}\left(\nabla\mathbf{u} + (\nabla\mathbf{u})^{\mathrm{T}}\right) \end{aligned} \tag{8}$$

with $\mathbb{C}$, a forth order material tensor and $\boldsymbol{\epsilon}$, the strain. Superscript T means the transpose of matrix. The deformation problem can be considered as a boundary value problem with boundary conditions given by

$$\boldsymbol{\sigma} : \mathbf{n} = \boldsymbol{t} \quad \text{or} \quad \mathbf{u} = \mathbf{u}_{\Gamma}\,, \quad \forall\, \boldsymbol{x} \in \partial\Omega\,. \tag{9}$$

3 Sub-Domain Finite Element Method

The sub-domain FEM consists of two steps (1) derivation of the weak FE formulations (Sect. 3.1) and (2) the sub-domain assembly (Sect. 3.2).

3.1 Weak forms

The method of weighted residuals is applied to derive the weak formulation of the all governing equations given in Sect. 2.

Assume $\mathcal{V}^{\mathrm{n}} \subset H^1_{\Gamma}(\Omega)^{\mathrm{n}}$ is the test function space. For all $w \in \mathcal{V}^1$, we have the weak form of the mass balance equation (1) as

$$\int_{\Omega} \left(\frac{\partial(nS^\gamma\rho^\gamma)}{\partial t} + \nabla\cdot\mathbf{q}^\gamma_f + \nabla\cdot(nS^\gamma\rho^\gamma\dot{\mathbf{u}}) - Q^\gamma_f \right) w\,\mathrm{d}\Omega = 0\,. \tag{10}$$

Applying integration by parts, equation (10) can be rewritten as

$$\begin{aligned} \int_{\Omega} \frac{\partial(nS^\gamma\rho^\gamma)}{\partial t} w\,\mathrm{d}\Omega - \int_{\Omega} \mathbf{q}^\gamma_f \cdot \nabla w\,\mathrm{d}\Omega + \int_{\Gamma} \mathbf{q}^\gamma_f \cdot \mathbf{n}\, w\,\mathrm{d}\Gamma \\ - \int_{\Omega} Q^\gamma_f w\,\mathrm{d}\Omega + \int_{\Omega} \nabla\cdot(nS^\gamma\rho^\gamma\dot{\mathbf{u}}) w\,\mathrm{d}\Omega = 0\,. \end{aligned} \tag{11}$$

Under the same assumption, the weak form of heat balance equation (4) can be obtained as

$$\begin{aligned} \int_{\Omega} \sum_{\gamma}^{phase} (\rho^\gamma C^\gamma_p) \frac{\partial T}{\partial t} w\,\mathrm{d}\Omega - \int_{\Omega} \mathbf{q}_{\mathrm{T}} \cdot \nabla w\,\mathrm{d}\Omega \\ + \int_{\Gamma} \mathbf{q}_{\mathrm{T}} \cdot \mathbf{n}\, w\,\mathrm{d}\Gamma - \int_{\Omega} Q^\gamma_{\mathrm{T}} w\,\mathrm{d}\Omega = 0\,. \end{aligned} \tag{12}$$

Taking account of nonlinearity, the weak form of the momentum balance equation (7) must be fulfilled throughout the load history, i.e.,

$$\int_{\Omega} \frac{1}{2}(\boldsymbol{\sigma} - \sum_{\gamma}^{phase} S^{\gamma} p^{\gamma} \mathbf{I}) : (\nabla \boldsymbol{w} + (\nabla \boldsymbol{w})^{\mathrm{T}})\, \mathrm{d}\Omega - \int_{\Omega} \boldsymbol{w}^{\mathrm{T}} \cdot \rho \mathbf{g}\, \mathrm{d}\Omega - \int_{\Gamma} \boldsymbol{w}^{\mathrm{T}} \cdot \mathbf{t}\, \mathrm{d}\Gamma = 0 \tag{13}$$

for all $\boldsymbol{w} \in \mathcal{V}^{\mathrm{n}}$, $\mathrm{n} = 2, 3$.

3.2 Sub-Domain Assembly

We use the Galerkin approach to deal with the above weak forms of balance equations, which are discretized in the finite element space. Assuming the mesh of domain Ω is discretized into m sub-domains, $\Omega = \Omega_0 \cup \Omega_1 \cup \cdots \cup \Omega_m$, local equation systems are assembled in sub-domains with sub-domain mesh topology and incorporating initial and boundary conditions. This leads to equation systems for sub-domains Ω_i

$$\begin{aligned} A_i^p p &= b_i^p \\ A_i^T T &= b_i^T \\ A_i^u u &= b_i^u \end{aligned} \tag{14}$$

where p, T and u are unknowns of fluid pressure, temperature and displacement, A_i^p, A_i^T and A_i^u are stiffness matrices, b_i^p, b_i^T and b_i^u are right hand sides of fluid flow process, heat transport process and deformation process, respectively. Since all variables are approximated by admissible finite element functions in the Taylor-Hood finite element space, i.e, low order interpolation for pressure and temperature variables and high order interpolation for displacement, respectively, the dimension of stiffness matrix A_i^u is n times of the dimension of A_i^p or A_i^T.

4 Parallelization of THM Coupled Processes

4.1 Parallelization Concept

The present study deals with the parallel finite element simulation of THM coupled problems. Basically, the principle computation time during a finite element simulation is spent in local/global assembly and solving linear equations arising from finite element discretization of bilinear forms of the corresponding IBVP (Sect. 2). As an inception study, the strategy of parallelization of the finite element analysis is aimed to parallelize local/global assembly and solving linear equations. The basic technique of such parallelization concept is to split the equation systems into a number of subsystems, cast the computation of subsystems to a number of processors (CPU nodes) concurrently and collect results from processors after all local computations are finished. Applying this technique to the FEM leads to a so called sub-structuring technique of domain decomposition (Farhat and Roux 1991, 1992). With the domain decomposition approach, sub-domains are discretized first from the original complete mesh. The mesh topology of those sub-domains has to be established. Then the splitting of the global equation systems is realized by performing global assembly in these sub-domains based sub-domain mesh topology. This distribution task is casted to processors concurrently. Up to this stage, we have split stiffness matrices and RHS vectors which are stored in the memory. The last step is to solve the linear equation system. For an iterative Krylov solver, the principle computations are multiplications of the system matrix by some vectors. Since the system matrix and right-hand-side (RHS) vector have been already split by sub-domains and since the system matrix is kept untouched during solver iterations, the computation of stiffness matrix multiplied by a vector is carried out by the means of computing multiplications of sub-domain matrices and vectors in the different processors concurrently. The communications among the processors to accumulate results of vector entries associated with element nodes at borders are related to sub-domains. It is assumed that all processes of a THM coupled problem share a unique mesh of domain. However, number of element nodes are different for thermal/hydraulic processes and deformation process, respectively. The so called Taylor-Hood finite element space has to be used due to the requirements of numerical stability and accuracy (Korsawe et al. 2006, Wang and Kolditz 2006). This is a specific problem of THM systems. A staggered (partitioned) scheme is applied to deal with the process couplings, i.e. the T/H/M processes are solved sub-sequentially in an iterative way.

4.2 Implementation of the Parallelization Algorithm

The implementation of the parallel finite element method is conducted with (MPI) within the framework of the open-source scientific software `GeoSys/RockFlow` (Kolditz et al. 2006).

We consider geometric parallelism, i.e., all involved processors run the same code for a parallel simulation and inter-processors communication is possible. Moreover, such simulations follow in a mainly procedure of reading data, constructing mesh topology of the whole domain, configuring variables for each process of THM coupled problems, allocating memory for sub-domain equations systems, constructing mesh topology of sub-domains, computing THM results at time steps and parallel processing of results output (Fig. 1). Within each time step, T, H and M processes are simulated individually embedded in a staggered coupling iteration. For such a scheme results are obtained by the classic procedure of assembling sub-domain equation system with initial and boundary conditions, solving linear equation system and probably going back to the assembling step if nonlinearity being considered. During the addressed procedure, three tasks in the simulation such as construction of sub-domain

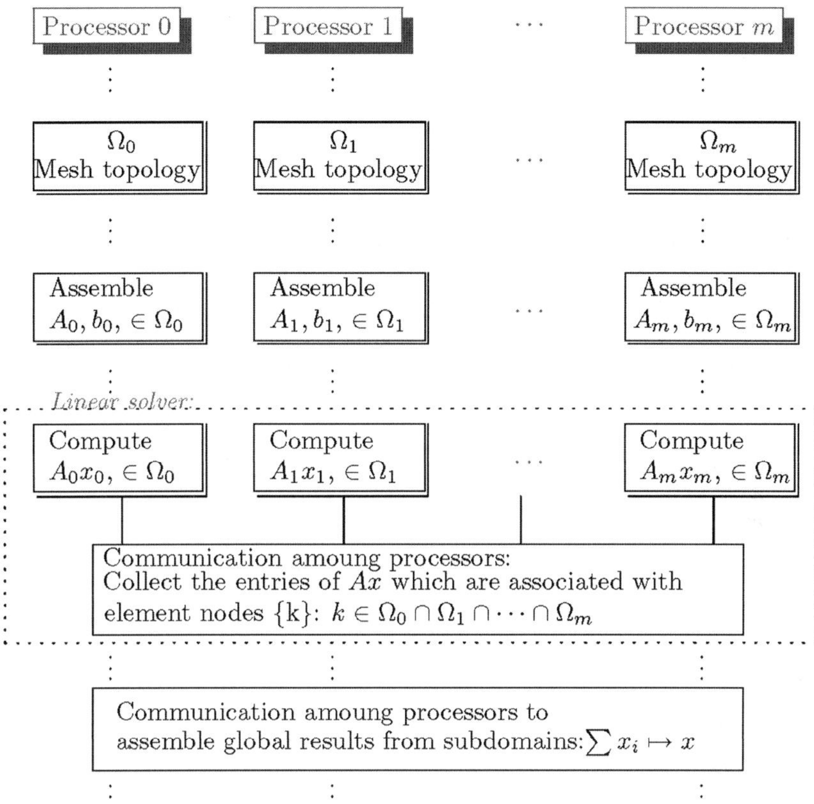

Fig. 1. Schematic of parallel finite element method

mesh topology, assembly of sub-domain equation systems and solving of the linear equations, can be distributed to the involved processors during a parallel simulation by sub-domains. The parallel performance of the first two tasks can be realized by just casting the computation in each sub-domain to different processors without any communication among processors. While, solution of linear equation systems can be obtained by performing a serial sub-domain matrix-vector multiplication in different processors and collecting multiplication results if entries of vectors are associated with element nodes at borders among sub-domains. As a prior condition to solve linear equation in such manner, Krylov space solver must be employed because that the multiplication of matrix and vector is its principle computation and the matrix itself kept unchanged. As for the linearized equations of THM processes, such sub-domain matrix-vector multiplication can be $A_i^p x_i^p$, $A_i^T x_i^T$ or $A_i^u x_i^u$. During the iteration steps of linear solver, communication among processors is required to update the solution. Figure 1 illustrates the procedure of parallel domain decomposition FE simulation. This parallelization is realized in MPI environment.

5 Test example

As a test example for THM coupled processes in an unsaturated porous medium, thermal biaxial test of a bentonite sample is analyzed. The geometry of the example is a rectangle with a size of 0.1×0.34 m. The domain is discretized into 340 quadrilateral elements with 385 nodes for the modeling of T and H processes and more than 385 nodes for the deformations analysis. Initial and boundary conditions are depicted in Fig. 2. Displacement load is prescribed on the top boundary increasing from 0 to 0.048 m within 100 time step with step size of 0.01 day. A segment of the bottom surface is assigned with constant temperature of 385°C. The whole domain is partially saturated with saturation of $S = 0.65$ and $T = 25°\mathrm{C}$ at the beginning of the simulation. Measured soil-water-characteristic curves are used in order to describe the thermo-hydraulic behaviour. Material properties and capillary pressure – as well as relative permeability – saturation functions, can be found in Wang and Kolditz (2006).

The example is simulated with 1, 2, 3, 4, 5 and 6 sub-domains, respectively, at a small Linux cluster. The number of involved processors is identical to the number of sub-domains for each simulation. The speedup is demonstrated in Fig. 3. The ratio of speedup vs number of processors is not constant. This is due to the time consumption by the communication among processors. A contour plot of results of saturation, temperature and vertical displacement is shown in Fig. 4.

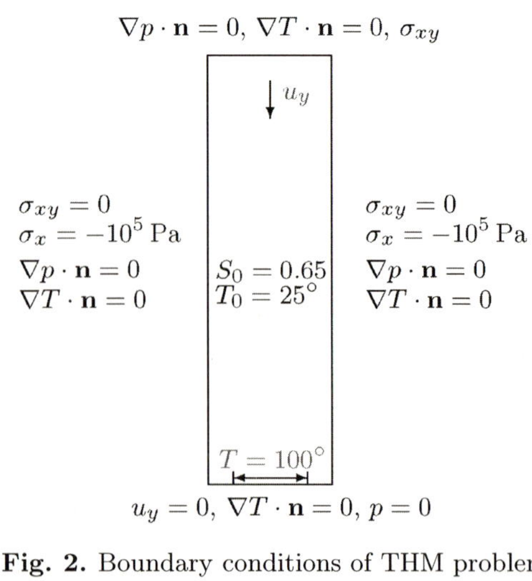

Fig. 2. Boundary conditions of THM problem

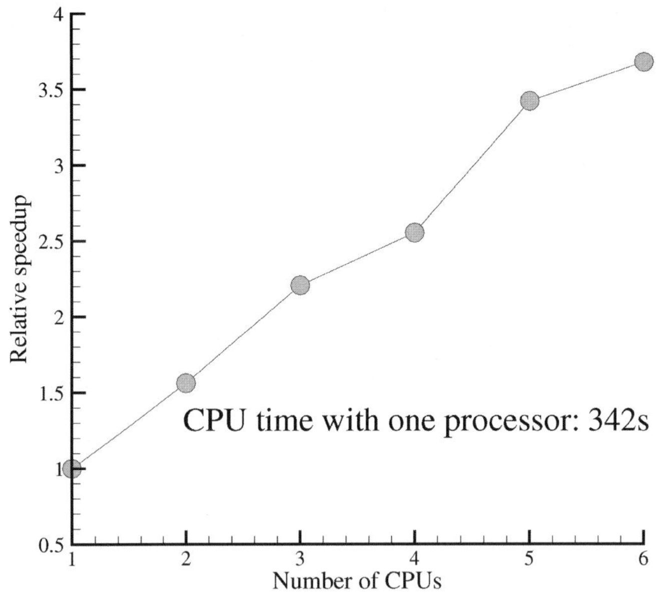

Fig. 3. Speedup of parallel FEM simulation of the THM test problem

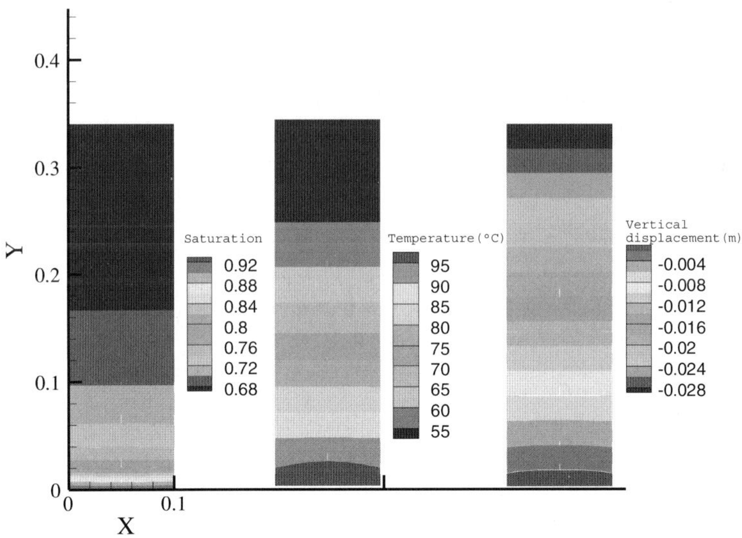

Fig. 4. Results of THM coupled simulation

6 Summary and Conclusions

A parallel finite element method (P-FEM) for the analysis of thermo-hydro-mechanical (THM) coupled processes in unsaturated porous media was developed and presented for a test example. The P-FEM is based on domain decomposition techniques. The global equation system of each THM process is split to sub-domains by the means of parallel sub-domain assembly of matrices and vectors. The iterative solver uses these sub-domain matrices and vectors directly to compute multiplication of matrix and vector to obtain the solution of the linear equation systems. The THM coupling is handled by the staggered scheme. The implementation of the P-FEM was done with MPI support. A THM test case is presented which was computed at a small Linux (6 CPU nodes) cluster. To our knowledge, the presented results are the first speedup measurements for THM coupled problems in partially saturated porous media. The achieved speedup has to be compared with results on larger clusters. This will be the next step for future work.

Acknowledgements

This work is funded by the Landesstiftung Baden-Württemberg in the priority program "Modelling and simulation at high performance computers" under grant 727 and by the German Ministry of Education and Research (BMBF) under grant 02C1117. We wish to thank R. Weber, H. Martini and Dr. Pitterich for their help in conducting this research.

References

de Boer R (2005) Trends in Continuum Mechanics of Porous Media: Theory and Applications of Transport in Porous Media. Springer-Verlag, Heidelberg

Ehlers W, Bluhm J (2002) Porous Media: Theory, Experiments and Numerical Applications. Springer-Verlag

Farhat C, Roux F (1991) A method of finite element tearing and interconnecting and its parallel solution algorithm, Int J Num Meth Eng 32(6):1205–1227

Farhat C, Roux F (1992) An unconventional domain decomposition method for an efficient parallel solution of large-scale finite element systems, SIAM J Sci Stat Comp 13(1):379–396

Fujisawa T, Inaba M, Yagawa G (2003) Parallel computing of high-speed compressible flows using a node-based finite-element method, Int J Num Meth Eng 58(3):481–511

Kolditz O (2002) Computational Methods in Environmental Fluid Mechanics. Springer, Berlin

Kolditz O, Wang W, de Jonge J, Xie M, Bauer S (2004) A process-oriented approach to compute thm problems in porous media. Part 1: Theoretical and informatics background. In: Schanz T (ed) Unsaturated Soils: Numerical and Theoretical Approaches. Springer, pp 53–66

Kolditz O, Xie M, Kalbacher T, Bauer S, Wang W, McDermott C, Chen C, Beyer C, Gronewold J, Kemmler D, Walsh R, Park CH, Du Y, Adamidis P, Hess M

(2006) GeoSys Rockflow, Open Source Software Design, Technical Report, Version 4.4 http://www.uni-tuebingen.de/zag/geohydrology. Technical report, Center for Applied Geosciences, University of Tuebingen

Korsawe J, Starke G, Wang W, Kolditz O (2006) Finite element analysis of poroelastic consolidation in porous media: mixed and standard approaches, Comp Meth App Mech Eng 195(9–12):1096–1115

Lewis RW, Schrefler BA (1982) A finite element simulation of the subsidence of gas reservoirs undergoing a waterdrive. In: Gallagher RH, Norrie DH, Oden JT, Zienkiewicz OC (eds) Finite Element in Fluids. John Wiley pp 179–199

Lewis RW, Schrefler BA (1998) The Finite Element Method in the Static and Dynamic Deformation and Consolidation of Porous Media (Second Edition). Wiley

Salinger AG, Xiao Q, Zhou Y, Derby JJ (1994) Massively parallel finite element computations of three-dimensional, time-dependent, incompressible flows in materials processing systems, Comp Meth App Mech Eng 119(1–2):139–156

Schanz T (ed) (2004) Unsaturated Soils: Numerical and Theoretical Approaches. Springer-Verlag, Heidelberg

Shioya R, Yagawa G (2005) Large-scale parallel finite-element analysis using the internet: a performance study, Int J Num Meth Eng 63(2):218–230

Stephanson O, Hudson J, Jing L (2003) Proceedings of the GeoProc International Conference on Coupled T-H-M-C Process in Geo-Systems. The Royal Institute of Technology, Stockholm, Sweden

Tezduyar TE, Sameh A (2006) Parallel finite element computations in fluid mechanics, Comp Meth App Mech Eng 195(13–16):1872–1884

Topping BHV, Khan AI (1996) Parallel finite element computations. Saxe-Coburg Publications, Edinburgh

Wang W, Kolditz O (2006) Object-oriented finite element analysis of thermo-hydromechanical (THM) problems in porous media, Int J Num Meth Eng (in print), doi: doi.wiley.com/10.1002/nme.1770

Development of an Object-Oriented Parallel Finite Element Code for Unsaturated Soils

Youliang Zhang and Domenico Gallipoli

School of Engineering, Durham University, South Road, Durham DH1 3LE, UK
youliang.zhang@durham.ac.uk, domenico.gallipoli@durham.ac.uk

Summary. The paper describes the outline of an object-oriented parallel finite element code for unsaturated soils currently developed at Durham University. Parallel iterative solvers based on domain decomposition methods are employed for the solution of the global system of equation, such as the preconditioned conjugate gradient (PCG) method and GMRES. A generic `Element` class is defined, which is capable of simulating various geotechnical problems.

Key words: finite element method, coupled problems, parallel computing

Introduction

The analysis of geotechnical problems in unsaturated soils often requires fully coupled analyses involving interaction of three phases and strongly nonlinear material behavior. Besides, most geotechnical problems cannot be accurately studied in two dimensions and they need instead three-dimensional models. Three-dimensional analysis and the non linearity of the governing differential system increase the amount of computation beyond the capabilities of any single personal computer. Alternative approaches are therefore required such as the parallelization of algorithms by using a cluster of computers working simultaneously.

The finite element implementation of coupled differential systems often results in large codes which might become unmanageable if not well designed. The introduction of object-oriented programming (OOP) enables writing of large and complex codes in an efficient, reusable, and extensible way. In the last decade many finite element codes in a wide range of disciplines were written by using OOP languages like `C++`, `Java`, etc. (Dubois-Pelerin and Zimmermann 1993, McKenna 1997) while parallel frontal techniques were used (Thomas et al. 1988, Wang and Schrefler 1998) to reduce computational time in multi-physical problems. The reader may consult Mackerle (2004) for an extensive list of references on the application of OOP to finite elements.

This paper describes a parallel finite element code for coupled multi-physics in unsaturated soils developed by using Domain Decomposition Methods (DDM) and object-oriented techniques.

The Mathematical and Physical Problem

Unsaturated soil is a three-phase porous medium consisting of solid grains, pore liquid and pore gas. In this work it is assumed that the gas phase only contains air whereas the liquid phase contains liquid water and dissolved air. The coupled study of liquid flow, gas flow and mechanical equilibrium requires the statement of governing differential equations as well as constitutive relationships between physical variables and primary unknowns. Governing equations are given by the equations of water and air mass balance and mechanical equilibrium as follows:

$$\frac{\partial n S_w \rho_w}{\partial t} + \mathrm{div}(\rho_w \mathbf{v}_w) = 0\,, \tag{1}$$

$$\frac{\partial}{\partial t}[n\rho_g(1 - S_w + HS_w)] + \nabla \cdot [\rho_g(\mathbf{v}_g + H\mathbf{v}_w)] = 0, \tag{2}$$

$$(\sigma_{ij} - \delta_{ij}u_g)_{,j} + u_{g,i} + b_i = 0 \tag{3}$$

where n is porosity, S_w is water saturation, ρ_w and ρ_g are the water and gas density, respectively, $\mathbf{v}_w$ and $\mathbf{v}_g$ are the water and gas fluxes, respectively, H is Henry's constant, σ_{ij} is the total stress, u_g is pore gas pressure, δ_{ij} is the Kronecker delta symbol and b_i is the body force.

After spatial discretization by Galerkin finite element method, the above equations are reduced to the algebraic system of Eq. (4), whose matrix and vector coefficients are discussed in detail in Gens et al. (1995). The primary unknowns of pore water pressure u_w, pore gas pressure u_g and displacement $\mathbf{u}$ appear in all equations making the problem fully coupled. This system is non-linear and the matrix coefficients depend on the primary unknowns through constitutive equations, e.g. the degree of saturation included in the coefficient matrix of Eq. (4) depends on displacement and suction. In summary, Eq. (4) constitutes the discretized form of a system of three non-linear partial differential equations, which must be simultaneously solved for the three primary unknowns in conjunction with appropriate constitutive relationships.

$$\begin{vmatrix} 0 & 0 & 0 \\ 0 & K_{gg} & K_{gw} \\ 0 & 0 & K_{ww} \end{vmatrix} \begin{vmatrix} \overline{u} \\ \overline{u}_g \\ \overline{u}_w \end{vmatrix} + \begin{vmatrix} C_{uu} & C_{ug} & C_{uw} \\ C_{gu} & C_{gg} & C_{gw} \\ C_{wu} & C_{wg} & C_{ww} \end{vmatrix} \begin{vmatrix} \frac{\partial \overline{u}}{\partial t} \\ \frac{\partial \overline{u}_g}{\partial t} \\ \frac{\partial \overline{u}_w}{\partial t} \end{vmatrix} = \begin{vmatrix} \frac{\partial f_u}{\partial t} \\ f_g \\ f_w \end{vmatrix} \tag{4}$$

where u_w, u_g, and u are nodal pore water pressure, nodal pore gas pressure and nodal displacement, respectively.

For simplicity, Eq. (4) can be written in a concise matrix form as:

$$\mathbf{KX} + \mathbf{C}\frac{\partial \mathbf{X}}{\partial t} = \mathbf{F}\,. \tag{5}$$

The temporal discretization of the above differential equations can be accomplished through the finite difference method. In this code an implicit difference method is used whereby Eq. (5) is approximated as:

$$\left\{\theta\mathbf{K} + \frac{1}{\Delta t}\mathbf{C}\right\}\mathbf{X}_{n+1} + \left\{(1-\theta)\mathbf{K} - \frac{1}{\Delta t}\mathbf{C}\right\}\mathbf{X}_n = \theta\mathbf{F}_{n+1} + (1-\theta)\mathbf{F}_n \tag{6}$$

where $0 \leq \theta \leq 1$. For non-linear problem, the coefficients matrices $\mathbf{K}$, $\mathbf{C}$ and vector $\mathbf{F}$ of the system equations depend strongly on the primary unknowns $\mathbf{X}$ and the solution vector $\mathbf{X}_{n+1}$ is obtained by iterative procedures like Picard and Newton–Raphson algorithms.

The Class Structure and Domain Decomposition Method

The proposed finite element code can be run on both sequential and distributed memory computers. The current version of the code is limited to hydro-mechanical coupled analyses (with two phases flow) but the classes are deliberately designed as flexible and general as possible to allow possible extensions in the future. The code can then be easily extended to carry out additional types of analyses such as, for example, the analyses of uncoupled mechanical problem, uncoupled seepage problem, coupled hydro-mechanical problems with single-phase flow and coupled thermo-hydro-mechanical problems with both single and multi-phase flow.

All classes are divided into four groups: preprocessing and post processing classes, finite element model classes, analysis and numerical classes and complementary classes. The second group encompasses `FEMModel`, `SubDomain` and its component classes including `Element`, `Node`, `Dof`, `Load`, `Material`, etc. This is similar to most OOP finite element codes.

Domain Decomposition Methods (DDM) are utilized as a "divide and conquer" strategy to achieve high performance computing on distributed memory parallel computers or networked workstations communicating via the Message Passing Interface (MPI) library. In this study, DDM are adopted together with some iterative solvers including PCG and GMRES (Saad and Schultz 1986) for symmetric and nonsymmetrical linear systems, respectively. The basic idea for DDM is that the physical domain of the finite element model is divided into a number of non-overlapping sub-domains mapped onto a different processor of a parallel computer. Computational tasks such as the calculations of stiffness matrix and external force in each sub-domain are accomplished by parallel processors, which also assemble the corresponding local system of equations. The distributed global system is then solved by iterative solvers with the help of communication routines among sub-domains. The DDM algorithm is also particularly useful for the solution of large finite element models requiring

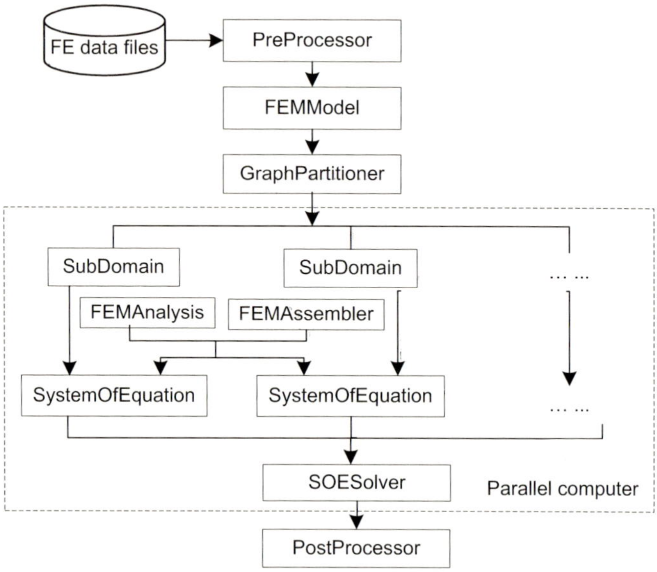

Fig. 1. Some top level classes and computational flow

substantial computer memory because each processor uses its own memory to store the variables of the corresponding sub-domain. The step-by-step DDM algorithm for finite element methods is illustrated in Fig. 1 and described as follows:

1. The finite element model input data are read by the `PreProcessor` object of the master processor, which builds the finite element model by using the `FEMModel` object. The master processor is responsible for administrating and coordinating the whole computational process.
2. `FEMModel` calls methods in the `GraphPartitioner` class to partition the global mesh into sub-domains that are distributed to different processors of parallel computers. The graph partition algorithm used in this work is "Metis" proposed by Karypis and Kumar (1998).
3. At each time step, the `FEMAssembler` object assembles the elemental and nodal contributions to generate the local system of equations in each sub-domain according to the type of problem and analysis method chosen. For a nonlinear system, the solution is iteratively by using a class derived from `FEMAnalysis` such as `PicardSolver` or `NewtonRaphsonSolver`.
4. The distributed global system of equation within each iteration is solved by means of the iterative solver such a PCG or GMRES based on the exchange of information between sub-domain boundaries in parallel. The main calculations inside the the iterative process are matrix-vector multiplication, dot product, and vector update, which are easy to parallelize. The `DistributedDataManager` class is used to store information about

neighbouring sub-domain boundaries, so that each sub-domain knows how to and where to send the relevant data.

5. Results from slave processors are transferred to the master processor where they are handled for visualization by the `PostProcessor` object.

`SubDomain` Class and Its Components

The `SubDomain` class is the kernel of the finite element model. The main functionalities of `SubDomain` object is to maintain and provide efficient access to component objects such as `Element`, `Node`, `Material`, `Load`, and `Dof`. The relationship among these classes is shown in Fig. 2 indicating that `Element` objects are associated to `Dof` objects, `Node` objects, `Material` objects, and `Load` objects. The standard UML signs are used in the following figures. The `Element` object returns the array of `Dof` location, geometrical data, load, and material information. The design of `Element` class is essential in achieving the goal of undertaking various types of geotechnical analyses in one code and it will be further discussed in the following section.

The `Material` class is introduced to provide a general interface handling different constitutive models for elasto-plastic materials (e.g. Mohr–Coulomb, Drucker–Prager and some advanced constitutive model for unsaturated soils). The `Load` class implements the methods calculating the values of loads at a given time step. The `Node` class obtains the calculated nodal loads from the `Load` class and passes the necessary information to the solver. `Node` objects have various degrees of freedom attached to and each `Dof` object has a type (i.e. displacement, pore water pressure, pore air pressure or temperature).

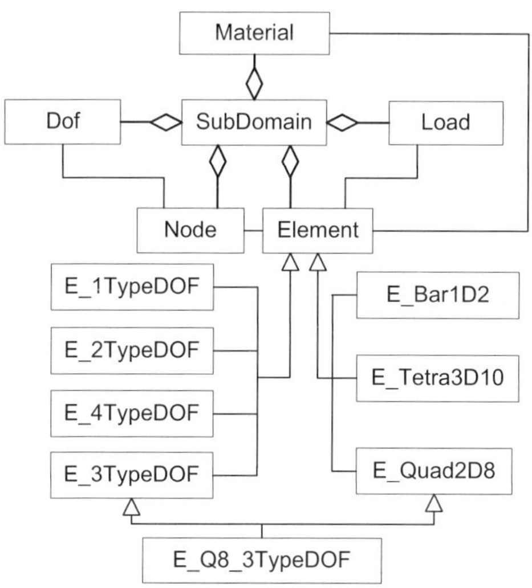

Fig. 2. Class SubDomain and its components

Element Class

The Element class is the base class for all types of elements. Each element object has a unique element number and contains a list of associated nodes, material, array of GaussPoint and array of Dof location. The Element objects are also associated with ShapeFunction objects which hold the geometric interpolation of the element. Different shape functions can be chosen for different primary unknowns and therefore more than one ShapeFunction object can be associated to an element.

Computational tasks for elements can be divided into geometrical computations and mechanical computations. Geometrical computations such as the calculation of strain-displacement matrix $\mathbf{B}$, ∇N_w (N_w is the shape function for pore water pressure) can be accomplished by calling the corresponding ShapeFunction objects. The mechanical computations involving the integration of the element matrices and external force are problem-dependent. For example, only the stiffness matrix needs to be calculated for uncoupled mechanical problem but additional matrices need to be calculated for coupled problems. To provide a generic interface for the Element class, the element stiffness matrix $\mathbf{K}_e$, capacity matrix $\mathbf{C}_e$, and external force vector $\mathbf{F}_e$ are defined as follows depending on the problem considered:

1. For uncoupled mechanical problem, the conventional forms of such matrices are used.
2. For the coupled hydro-mechanical problem of Eq. (4), the matrices are:

$$\mathbf{K}_e = \begin{vmatrix} 0 & 0 & 0 \\ 0 & K_{gg} & K_{gw} \\ 0 & 0 & K_{ww} \end{vmatrix}, \quad \mathbf{C}_e = \begin{vmatrix} C_{uu} & C_{ug} & C_{uw} \\ C_{gu} & C_{gg} & C_{gw} \\ C_{wu} & C_{wg} & C_{ww} \end{vmatrix}, \quad \mathbf{F}_e = \begin{vmatrix} \mathbf{F}_u \\ \mathbf{F}_g \\ \mathbf{F}_w \end{vmatrix}. \tag{7}$$

3. For thermo-hydro-mechanical problem:

$$\mathbf{K}_e = \begin{vmatrix} K_{uu} & K_{ug} & K_{uw} & K_{uT} \\ K_{gu} & K_{gg} & K_{gw} & K_{gT} \\ K_{wu} & K_{wg} & K_{ww} & K_{wT} \\ K_{Tu} & K_{Tg} & K_{Tw} & K_{TT} \end{vmatrix}, \quad \mathbf{C}_e = \begin{vmatrix} C_{uu} & C_{ug} & C_{uw} & C_{uT} \\ C_{gu} & C_{gg} & C_{gw} & C_{gT} \\ C_{wu} & C_{wg} & C_{ww} & C_{wT} \\ C_{Tu} & C_{Tg} & C_{Tw} & C_{TT} \end{vmatrix}. \tag{8}$$

Each derived Element class must provide methods to compute these matrices. The hierarchy of the Element class is illustrated in Fig. 2 where the classes E_Bar1D2 (2-node 1D bar element), E_Quad2D8 (8-node 2D quadrilateral element) and E_Tetra3D10 (10-node 3D tetrahedral element) are used for the elemental geometrical computations. For computations of elemental contributions depending on specific problem type, the classes E_1TypeDOF (uncoupled mechanical problem or single-phase seepage problem), E_2TypeDOF (coupled hydro-mechanical problem for saturated soils), E_3TypeDOF (coupled hydro-mechanical problem for unsaturated soils) and E_4TypeDOF (coupled thermo-hydro-mechanical problem in unsaturated soils) are introduced. It is then natural to derive a new element class by using multiple inheritances. For example, a two dimensional 8-node element for hydro-mechanical problem in unsaturated soils (`E_Q8_3TypeDOF`) is derived from `E_Quad8` and `E_3TypeDOF`.

Conclusions

The outline of an object-oriented parallel finite element code for unsaturated soils is presented. The flexible structure of the code allows for easy implementation of different categories of multi-physical geotechnical problems by deriving new element and analysis classes based on existing classes. The Domain Decomposition Method together with iterative solvers are employed to parallelize the code. New solvers for symmetric and non-symmetric equations can also be added with little modifications of the existing code. The input data for parallel and serial computations are the same as mesh partition is carried out inside the code. Some other classes are not described here due to page limitations.

References

Dubois-Pelerin Y, Zimmermann T (1993) Object-oriented finite element programming: III An efficient implementation in C++, Computer Meth Appl Mech Eng 108:165–183

Gens A, Jouanna P, Schrefler BA (1995) Modern issues in non-saturated soils. CISM course and Lectures No. 357. Springer-Verlag, New York

Karypis G, Kumar V (1998) Multilevel k-way partitioning scheme for irregular graphs, J Parallel and Distributed Computing 48:96–129

Mackerle J (2004) Object-oriented programming in FEM and BEM a bibliography (1990–2003), Advances in Engineering Software 35:325–336

McKenna FT (1997) Object Oriented Finite Element Programming: Framework for Analysis, Algorithms and Parallel Computing. Ph.D. Thesis, University of California

Saad Y, Schultz MH (1986) GMRES: A generalized minimal residual algorithm for solving non-symmetric linear systems, SIAM J Scientific Statist Comput 7:856–869

The Message Passing Interface (MPI) standard. http://www-unix.mcs.anl.gov/mpi/

Thomas HR, Yang HT, He Y, Jefferson AD (1998) Solving coupled thermo-hydro-mechanical problems in unsaturated soil using a substructuring Frontal technique, Communications in Numerical Methods in Engineering 14:783–792

Wang X, Schrefler BA (1998) A multi-frontal parallel algorithm for coupled thero-hydro-mechanical analysis of deforming porous media, Int J Numer Meth Engng 43:1069–1083

Part III

Applications

Modeling Unsaturated Flow and Atmospheric Interactions

Craig H. Benson

Geo Engineering Program, University of Wisconsin, Madison, WI 53706, USA
benson@engr.wisc.edu, www.uwgeoengineering.org

Summary. Four codes used to simulate unsaturated flow with atmospheric interactions are described and evaluated. The codes are similar conceptually and function in a similar manner. However, the codes use different algorithms, and therefore yield different predictions for the same input. A field validation exercise conducted with three of the codes shows that the predictions appear realistic regardless of which code is being used. However, each code provides a different prediction, and none of the predictions are in agreement with field measurements. The differences can be attributed in part to the methods used to implement the atmospheric boundary. More subtle features of the codes also have a significant effect on predictions (e.g., hydraulic properties model, hysteresis, vapor flow, thermally-driven flow, transpiration algorithms, ground freezing, snow melt, etc.). The user should consider each of these features carefully so that realistic, but conservative predictions are obtained.

Key words: unsaturated flow, atmospheric boundary, numerical model, percolation, run off, hydraulic properties, pore interaction term

Introduction

Flow in the unsaturated zone and atmospheric interactions play an important role in many geoenvironmental engineering problems. Examples include landfill covers where water balance principles are used to limit percolation into underlying waste, drainage and desiccation of tailings deposited by aqueous methods in mine waste impoundments, and groundwater recharge in response to infiltration. In each of these examples, the behavior of the engineered or natural system is affected or controlled by unsaturated flow, and the response of flow processes to atmospheric conditions.

Unsaturated flow is a highly non-linear process, which complicates analysis of the unsaturated zone. Consequently, engineers traditionally have used simplified solutions for analysis of unsaturated flow problems (e.g., the Green–Ampt infiltration model is a simplified solution of the non-linear infiltration process). However, affordable high-speed computers and powerful codes with graphical user interfaces now make direct solutions of non-linear unsaturated flow problems tractable, even for the practicing engineer (Benson 2006).

This paper describes four codes that are commercially available and commonly used for modeling unsaturated flow with atmospheric interactions. Predictions made with the codes are compared to field data, and selected issues affecting the accuracy of the model predictions are described.

Simulating Unsaturated Flow

Richards' equation is the governing partial differential equation for unsaturated flow:

$$\frac{\partial \theta}{\partial t} = \nabla \bullet [K_\psi \nabla (z - \psi)] - S \tag{1}$$

where θ is volumetric water content, t is time, K_ψ is hydraulic conductivity, ψ is matric suction, z is the vertical coordinate, and S is a sink term that is frequently used to simulate the uptake of water by roots within the unsaturated zone. The derivation of Eq. (1) assumes isothermal conditions, isotropic hydraulic conductivity, an incompressible water phase and pore space, and a static and continuous vapor phase that has spatially uniform atmospheric pressure.

Solution of Eq. (1) usually involves rewriting the temporal derivative of θ in terms of the temporal derivative of ψ:

$$C_\psi \frac{\partial \psi}{\partial t} = \nabla \bullet [K_\psi \nabla (z - \psi)] - S \tag{2}$$

where $C_\psi = \partial\theta/\partial\psi$ (i.e., the slope of the soil–water characteristic curve, or SWCC) and is referred to as the specific water capacity. The specific water capacity varies between $-\infty$ and 0, and can be very negative for soils with very uniform pore size distribution. The non-linearity of Eq. (2) arises because K_ψ is a function of ψ. Moreover, as illustrated in Fig. 1, both C_ψ and K_ψ vary over a large range and in a highly non-linear manner with ψ, greatly complicating solution of Eq. (2). Hysteresis in K_ψ and c_ψ further complicate the solution.

Codes to Solve Richards' Equation

A variety of powerful codes are available for solving Eq. (2). The most commonly used codes, `HYDRUS`, `SVFLUX`, `UNSAT-H`, and `VADOSE/W`, are listed in Table 1. Each of these codes includes an atmospheric boundary to simulate atmospheric interactions and root–water uptake functions to simulate plant transpiration. Factors such as solute transport, heat transfer and thermally driven flow, and vapor flow are also incorporated in these codes using modified versions of Eq. (2) and coupled solutions. Each of these codes is available with a graphical user interface and runs in the Windows™ operating system.

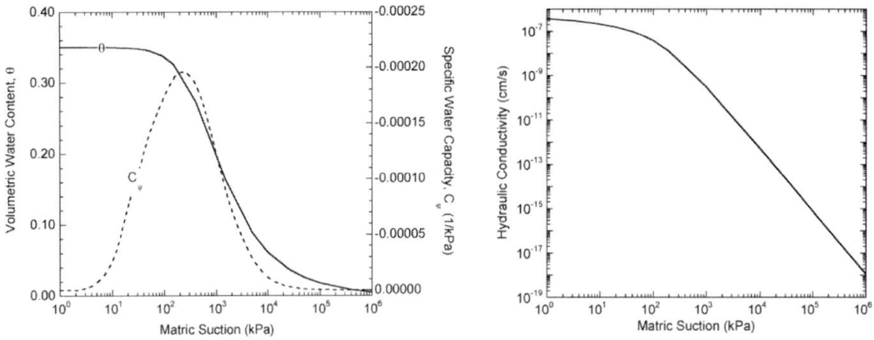

Fig. 1. Soil water characteristic curve (SWCC), specific water capacity, and hydraulic conductivity as a function of matric suction

Table 1. Codes commonly for unsaturated flow modeling with atmospheric interactions

Code	Source	Dimensionality	Other Features
HYDRUS	pc-progress.com	1, 2, or 3D	Solute and colloid transport, heat transfer, dual porosity, hysteresis, snow hydrology, runoff, stochastic soil properties
SVFLUX	soilvision.com	1, 2, or 3D	Heat transfer, ground freezing, stochastic soil properties, runoff
UNSAT-H	hydrology.pnl.gov or uwgeosoft.org	1D	Vapor flow, heat transfer, thermally driven flow, run off, hysteresis
VADOSE/W	geoslope.com	1D or 2D	Oxygen transport, snow hydrology, ground freezing, run off and down slope infiltration, heat transfer, volume change (limited), vapor flow

The newer codes, HYDRUS, SVFLUX, and VADOSE/W, employ the finite element method to solve Richards' equation. UNSAT-H uses the finite-difference method, and tends to be more stable when solving problems where large contrasts in hydraulic properties exist at layer interfaces (e.g., capillary barriers).

Boundary Conditions

Each of these codes includes a variety of boundary conditions to simulate various problems. A flux boundary condition is generally applied at the surface to simulate atmospheric interactions. Boundary conditions generally applied at the bottom of the profile include a unit gradient condition (e.g., gravity-driven flux = K_ψ at the bottom of the domain), a seepage face (e.g., flux = the saturated hydraulic conductivity, K_s, when the boundary is saturated; otherwise flux = 0), a prescribed head, or a prescribed flux. For 2D and 3D simulations, prescribed head and prescribed flux boundaries are often applied along the sides of the domain.

Conceptually, a similar atmospheric flux boundary condition is used in each of the codes. Infiltration occurs during precipitation at a rate governed by the hydraulic properties of the profile (i.e., the "infiltration capacity" approach), and precipitation exceeding the infiltration capacity is assumed to be run off. Evaporation is assumed to occur from the soil surface and is bounded by the potential evaporation (PE) rate. The specific methods by which these mechanisms are implemented, however, vary from code to code. For example, some codes set the evaporative flux (E) to zero during precipitation, whereas others apply a net precipitation that is equal to the precipitation rate less PE during the precipitation event. Differences also exist between the methods used to compute E. For example some codes compute E based on Fick's law and differences between the water vapor pressure in the atmosphere and the pore space at the ground surface, whereas others use empirical relationships relating actual E to PE and the soil suction at the surface. A description of the different methods used to implement the atmospheric boundary can be found in Scanlon et al. (2002).

Verification and Validation

Verification and validation are essential steps to demonstrate the validity of any numerical code. Verification is a direct comparison between predictions made with a numerical code and known solutions. Validation is a comparison between predictions made with a numerical code and measured behavior. Verification is conducted by nearly all code developers to demonstrate that the numerical solution is correct (e.g., see the verification tests in Fayer (2000) for `UNSAT-H`). Validation is less common, particularly for complex problems that involve time varying boundary conditions such as atmospheric interactions.

A key requirement for validation is to input measured quantities for all variables that have a significant effect on the solution. Favorable comparisons between measurements and model predictions obtained from simulations where key input variables are estimated are more akin to model calibration exercises than validation exercises. This is a particularly significant problem for unsaturated flow models with atmospheric interactions, because few data sets exist where all of the key input variables and responses have been measured a priori.

Validation Exercise

A validation exercise was conducted with `HYDRUS`, `UNSAT-H`, and `VADOSE/W` using data collected from a field site monitored as part of the US EPA's Alternative Cover Assessment Program (ACAP) (Albright et al. 2004). `SVFLUX` was not available for the validation exercise. The field site is in Altamont, California (CA), USA, which is approximately 80 km East of San Francisco. The site is semi-arid with an average annual precipitation (P_a) = 358 mm and an average annual ratio of precipitation to potential evapotranspiration (P/PET_a) = 0.31. All of the precipitation at the site occurs as rain; snow and frozen ground are extremely rare.

Test Section and Monitoring Data

Data were collected from an instrumented test section used to monitor the performance of a water balance cover planned for a municipal solid waste landfill at the site. The cover was a monolithic barrier consisting of a 1060 mm thick storage layer overlying 150 mm of interim cover soil (Fig. 2) that was placed on a 5% slope. Crushed claystone available on site was used to construct the storage layer and the interim cover soil. The monitoring system consisted of a 10 m × 20 m lysimeter used to collect percolation from the base of the cover, a runoff collection system, three nests of sensors to measure water content and matric suction within the cover profile, and a weather station to measure meteorological conditions on site. A detailed description of the test section and instrumentation can be found in Benson et al. (1999, 2001). Bohnhoff (2005) contains a compilation of the field data.

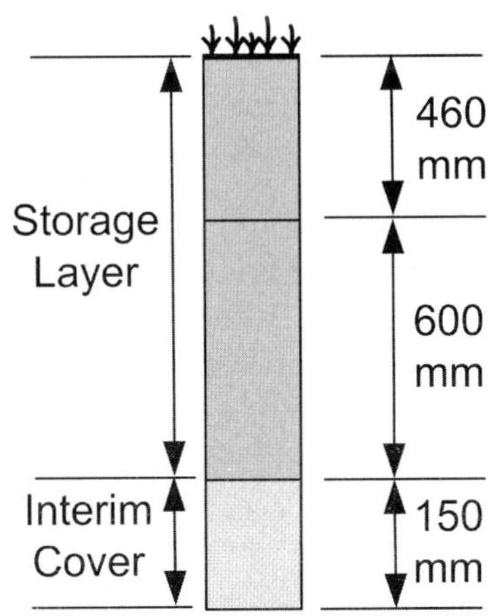

Fig. 2. Profile of water balance cover at ACAP site in Altamont, CA

Soil and Vegetation Properties

A collection of samples were collected as undisturbed blocks during construction to characterize the hydraulic properties of the cover soils. Specimens trimmed from the blocks were tested to determine the saturated hydraulic conductivity and the drying soil water characteristic curve. The SWCC was parameterized with van Genuchten's equation and the unsaturated hydraulic conductivity was computed using the van Genuchten-Mualem equation (van Genuchten 1980) with a pore interaction term $(\ell) = 0.5$.

Mean hydraulic properties were assigned to three parts of the cover profile: upper 460 mm of the storage layer, lower 600 mm of the storage layer, and the interim cover soil. The means are summarized in Table 2 (geometric means for K_s and van Genuchten's α parameter; arithmetic means for the saturated water content, θ_s, residual water content, θ_r, and van Genuchten's n parameter). Samples were also collected from the upper 460 mm in 2002 and 2003 to obtain hydraulic properties of the near surface soil as the cover profile weathered. Saturated hydraulic conductivities obtained from these samples are summarized in Table 2.

Table 2. Summary of mean hydraulic properties for Altamont test section

Layer	Thickness (mm)	SWCC Parameters				Saturated Hydraulic Conductivity (cm/s)		
		θ_r	θ_s	α (1/m)	n	2001	2002	2003
Upper portion of storage layer	460	0.00	0.37	0.050	1.33	5.3×10^{-7}	2.2×10^{-6}	1.1×10^{-4}
Lower portion of storage layer	600	0.00	0.35	0.025	1.54		4.5×10^{-7}	
Interim Cover	300	0.00	0.35	0.057	1.36		3.0×10^{-6}	

The cover profile was seeded with a mixture of local grasses. Surveys were conducted periodically to determine the leaf area index (LAI) and samples were collected for measurement of the distribution of root density. A summary of the LAI and root density data can be found in Bohnhoff (2005).

Simulation Results

Predictions of the water balance for the test section at Altamont were made using `HYDRUS`, `UNSAT-H`, and `VADOSE/W` with identical input. Daily meteorological data were used and hysteresis in the hydraulic properties was ignored. An atmospheric boundary was used at the surface and a unit gradient boundary was used at the base of the profile. All simulations were conducted in 1D. A detailed description of the input can be found in Bohnhoff (2005).

A comparison of the predictions is shown in Fig. 3 in terms of cumulative water balance quantities. The predictions appear very realistic, with responses that reflect the temporal variations in meteorological conditions. Each code also provides a different prediction, even though all of the codes solve the same basic partial differential equation (Eq. 3) and were supplied with identical input. For example, runoff predicted by `HYDRUS` and `VADOSE/W` is slightly lower than runoff measured in the field, whereas `UNSAT-H` grossly over-predicts run off. `UNSAT-H` under-predicts evapotranspiration (ET) significantly, whereas `HYDRUS` and `VADOSE/W` over-predict ET. Soil water storage predicted by `VADOSE/W` is reasonably close to the measured soil water storage. `HYDRUS` tends to under-predict soil water storage, particularly during the drier

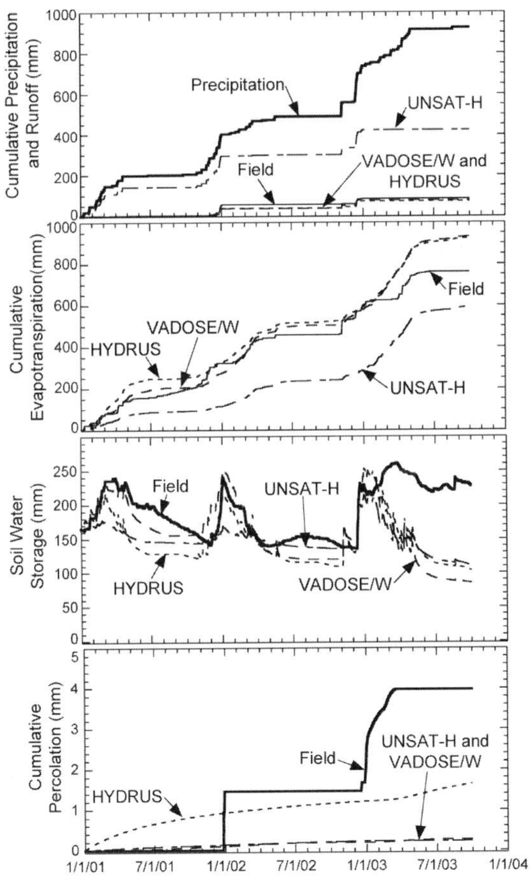

Fig. 3. Water balance data from ACAP test section in Altamont, CA and predictions made with `HYDRUS`, `UNSAT-H`, and `VADOSE/W`

periods. UNSAT-H also under-predicts soil water storage (more so than HYDRUS), except during the driest periods. Soil water storage predicted by UNSAT-H also lacks the seasonal fluctuations observed in the field and predicted by the other codes. None of the codes capture the absence of a seasonal drop in soil water storage at the end of the record, and all of the codes under-predict percolation.

Many of the differences in the predictions are related to how the atmospheric flux boundary condition is applied. For example, UNSAT-H over-predicts runoff because daily precipitation is applied at 10 mm/hr (the default), which is much higher than the intensity commonly observed in the field. As a result, the infiltration capacity is readily exceeded, and too much run off is predicted. As shown in Fig. 4, applying precipitation at the average intensity observed in the field results in less runoff, but does not resolve the over-prediction completely. Unlike UNSAT-H, HYDRUS and VADOSE/W apply daily precipitation uniformly throughout a 24 hr period. While this application method may not faithfully represent the intensity at which precipitation occurs in the field, the method does yield a more accurate prediction of runoff (Fig. 3). Gitirana et al. (2005) indicate that over-prediction of run off can be caused by poor approximation of large gradients in pore water pressure and material properties near the surface. They show that adaptive mesh refinement methods, such as those used in SVFLUX, improve these approximations and result in more accurate predictions of run off.

The over-estimation of ET by HYDRUS and VADOSE/W is also tied to the method used to simulate the atmospheric boundary. Both of these codes compute a net precipitation equal to the difference between actual precipitation and potential evaporation during the precipitation event. This approach over-estimates the amount of ET that is occurring, and under-estimates the amount of water contacting the surface (Scanlon et al. 2002).

The impact of the lower boundary condition has received considerable discussion in the literature (e.g. Scanlon et al. 2002, Benson et al. 2004), but generally has a smaller impact on predictions than the upper boundary condi-

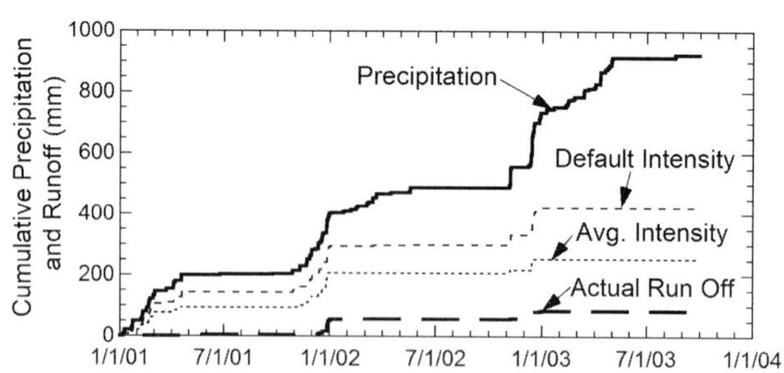

Fig. 4. Actual run off measured in the field and run off predicted by UNSAT-H for ACAP test section in Altamont, CA. Predictions made using default precipitation intensity (10 mm/hr) and average precipitation intensity observed in the field (0.68 mm/h)

tion. An example is shown in Fig. 5, where predictions were made with HYDRUS using the lower boundary set as a unit gradient condition or a seepage face condition. For the ACAP lysimeter being simulated, the seepage face condition provides a more accurate representation of the capillary break caused by the drainage layer at the based of the lysimeter. However, as shown in Fig. 5, both boundary conditions result in an under-prediction of percolation, and the under-prediction is larger when the seepage face boundary is used instead of the unit gradient boundary. Similar predictions were obtained with VADOSE/W with the lower boundary set as a unit gradient condition or a seepage face condition.

Hydraulic Property Functions

Constitutive Models

Most codes provide options for different constitutive models to describe the SWCC and the hydraulic conductivity function. In most cases, van Genuchten's function is used to describe the SWCC and the van Genuchten–Mualem function is used to describe the unsaturated hydraulic conductivity. These functions are commonly used because they are continuous and smoothly varying, and therefore easily fit to data and implemented in computer programs. However, their use may not result in the most realistic predictions of unsaturated flow problems with soil-atmosphere interactions, as illustrated in the following example.

Two simulations of the ACAP test section in Altamont, CA were conducted with UNSAT-H using the input data described previously. One of the simulations was conducted using van Genuchten's function to describe the SWCC and the van Genuchten–Mualem function to describe the unsaturated hydraulic conductivity. Another identical simulation was conducted using the Brooks–Corey functions to describe the SWCC and the unsaturated hydraulic

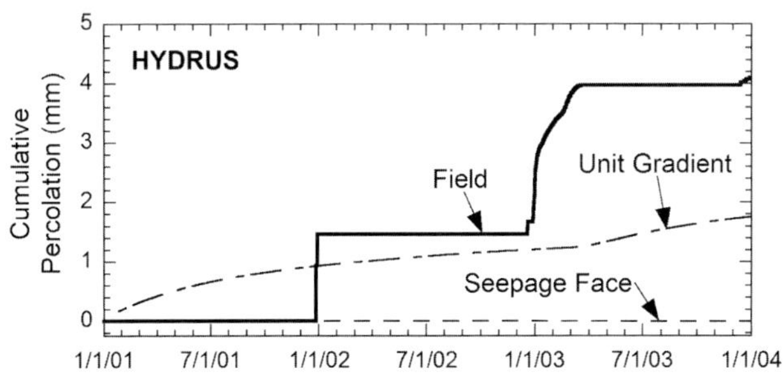

Fig. 5. Percolation measured in the field and percolation predicted by HYDRUS for ACAP test section in Altamont, CA with the lower boundary condition set as a unit gradient or as a seepage face condition

conductivity (Brooks and Corey 1966). The Brooks–Corey function was parameterized by fitting it to the van Genuchten function originally used for the SWCC. As shown in Fig. 6, the two SWCCs are virtually identical, except near the air entry suction (~ 100 kPa). In contrast, the hydraulic conductivity predicted by the van Genuchten–Mualem function is consistently lower than the hydraulic conductivity predicted by the Brooks–Corey function, especially near the air entry suction.

Predictions from `UNSAT-H` using the van Genuchten–Mualem and Brooks–Corey functions are shown in Fig. 7. Run off predicted using the Brooks–Corey function is in close agreement with run off measured in the field, and is much lower than run off predicted with the van Genuchten–Mualem function. Better agreement is obtained using the Brooks–Corey function because the function predicts higher hydraulic conductivities near saturation, which strongly affect infiltration.

The better agreement between predicted and measured run off obtained using the Brooks–Corey function affects the other water balance quantities too. Soil water storage predicted using the Brooks–Corey function is in much

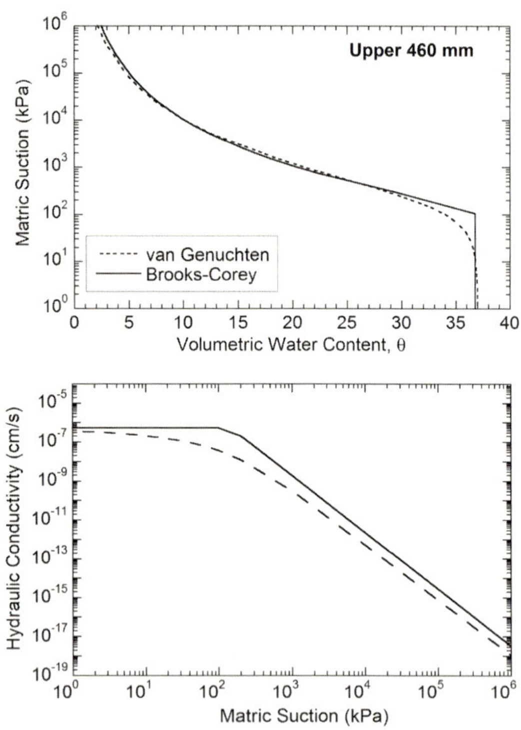

Fig. 6. Brooks–Corey, van Genuchten, and van Genuchten–Mualem functions used to describe the SWCC and unsaturated hydraulic conductivity of the upper 460 mm of soil for the ACAP test section in Altamont, CA

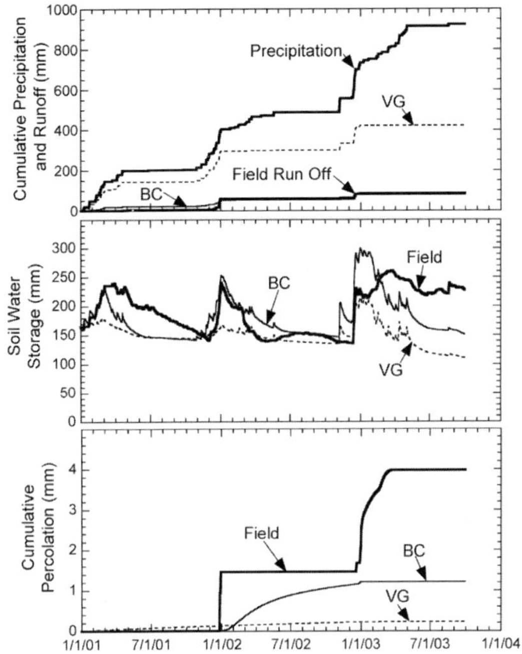

Fig. 7. Predicted and measured run off, soil water storage, and percolation for the ACAP test section in Altamont, CA. Predictions made with UNSAT-H using Brooks–Corey, van Genuchten, and van Genuchten–Mualem functions

better agreement with the field-measured soil water storage than the prediction obtained with the van Genuchten–Mualem function, particularly during the second year of monitoring (2002). Soil water storage predicted using the Brooks–Corey function also exhibits the seasonal fluctuations present in the field data. Similarly, percolation predicted using the Brooks–Corey function is closer to percolation measured in the field, although percolation is underpredicted using both the Brooks–Corey and van Genuchten–Mualem functions.

Pore Interaction Term

Another significant variable is the pore interaction term in the hydraulic conductivity function. For the van Genuchten–Mualem function, the pore interaction term (ℓ) is found in the denominator and affects the rate at which the hydraulic conductivity changes with suction:

$$K_\psi = K_s \frac{\{1-(\alpha\psi)^{n-1}[1+(\alpha\psi)^n]^{-m}\}^2}{[1+(\alpha\psi)^n]^{\ell/2}} \,. \tag{3}$$

In nearly all applications, the pore interaction term in the van Genuchten–Mualem function is assumed to be 0.5, as was originally suggested in van

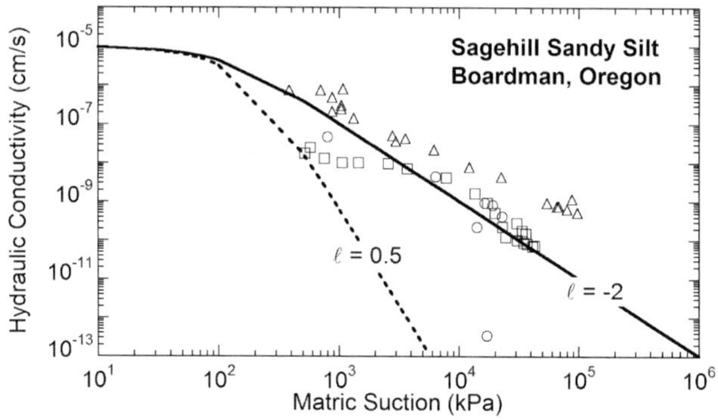

Fig. 8. Unsaturated hydraulic conductivity of Sagehill sandy silt from Boardman, Oregon. Smooth lines correspond to predictions made with the van Genuchten–Mualem function using $\ell = 0.5$ and -2. Data points are from three sets of measurements made using the instantaneous profile method described in Meerdink et al. (1995)

Genuchten (1980). However, more recent research has suggested that the pore interaction term varies with soil type (Schaap and Leij 2000). In particular, $\ell = 0.5$ is reasonable for clean coarse-grained materials, whereas $\ell = -2$ is more suitable for fine-textured soils.

As shown in Fig. 8, changing ℓ from 0.5 to -2 has a significant effect on the hydraulic conductivity function. At $\psi = 1000\,\text{kPa}$, for example, the hydraulic conductivity differs by two orders of magnitude for $\ell = 0.5$ and -2. This large difference in hydraulic conductivity has significant effects on predictions made with unsaturated flow codes, as illustrated in Fig. 9. Simulations of an ACAP test section in Sacramento, CA were made with UNSAT-H using ℓ ranging from 0.5 to -2. This test section is nearly identical to the test section in Altamont, CA and was constructed with similar methods. A detailed description of the test section can be found in Bohnhoff (2005). As shown in Fig. 9, decreasing ℓ resulted in larger fluctuations in soil water storage. Run off also decreased, and evapotranspiration and percolation increased (not shown). Each of these responses reflects that higher hydraulic conductivity is predicted at a given suction as ℓ decreases.

Dimensionality

Unsaturated flow problems with atmospheric interactions are often conducted in 2D or 3D when the soil surface is sloping or the layering in the profile promotes multi-dimensional flow. More realistic simulations can be conducted in 2D and 3D for such problems, but the analysis requires more computing time and memory. Consequently, 1D analyses are used as an approximation in many cases.

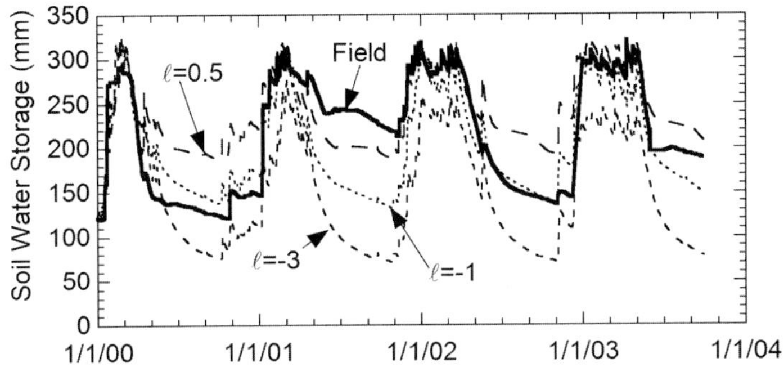

Fig. 9. Soil water storage measured in the field for ACAP test section in Sacramento, CA and predicted with UNSAT-H using $\ell = 0.5$, -1, and -3

The difference between 1D and 2D predictions was evaluated by simulating an ACAP test section located in Marina, CA using HYDRUS in the 1D and 2D modes. The test section in Marina has the same dimensions as the test section in Altamont, CA, but is on a 4H:1V slope. A description of the test section can be found in Albright et al. (2004). The 2D simulations were conducted in a domain representing the entire test section. The 1D simulations considered only a vertical profile through the test section without any slope.

Predictions of run off, soil water storage, and percolation obtained from HYDRUS in 1D and 2D are shown in Fig. 10. The 2D simulation was conducted over a shorter time period due to numerical problems that terminated the analysis. Less run off occurs and more soil water storage accumulates during wet periods when the simulation is conducted in 2D. Less percolation is also predicted in 2D than in 1D. However, the differences between the predictions are modest. Thus, at least in this case, a 1D simulation probably would be adequate, even though the actual problem is 2D or 3D.

Summary and Practical Implications

Four codes used to simulate unsaturated flow with atmospheric interactions have been described in this paper. Each uses numerical methods to solve Richards' equation using boundary conditions that reflect atmospheric fluxes. The codes include functions to account for other processes as well, such as root water uptake, heat transfer, vapor flow, and ground freezing. The codes are similar conceptually and function in a similar manner. However, the codes use different algorithms, and therefore yield different predictions for the same input.

A field validation exercise conducted with three of the codes has shown that the predictions appear realistic regardless of which code is being used. However, each code provides a different prediction, and none of the predictions are in agreement with the field measurements. These differences in the

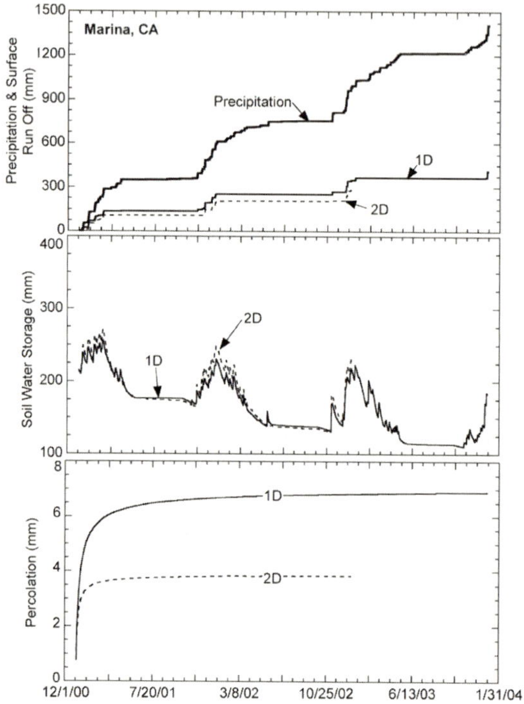

Fig. 10. Predictions from comparative simulations conducted in 1D and 2D using HYDRUS and input data from an ACAP test section in Marina, CA, USA

predictions can be attributed in part to differences in how the atmospheric boundary is implemented. Refining how this boundary should be implemented is a topic in need of research.

More subtle characteristics of the codes can also have a significant effect on the predictions. For example, the constitutive model used for the hydraulic properties can have a significant effect on predictions. Other processes may also affect predictions (e.g., hysteresis, vapor flow, thermally-driven flow, transpiration algorithms, ground freezing, snow melt, etc.). These processes have not been considered in this paper due to space limitations, but their importance should be considered in any modeling exercise.

Acknowledgements

Funding for this study has been provided by the US Environmental Protection Agency's (USEPA) Superfund Innovative Technology Evaluation (SITE) Program, the US Environmental Protection Agency's Science to Achieve Results (STAR) Program (Grant No. R-82951501-0, Rocky Mountain Hazardous

Substance Research Center), and the US National Science Foundation (Grant No. CMS-0625850).

References

Albright W, Benson C, Gee G, Roesler A, Abichou T, Apiwantragoon P, Lyles B, Rock S (2004) Field water balance of landfill final covers, J Environ Quality 33(6):2317–2332

Benson C, Abichou T, Wang X, Gee G, Albright W (1999) Test Section Installation Instructions – Alternative Cover Assessment Program. Environmental Geotechnics Report 99–3. Dept of Civil & Environmental Engineering, University of Wisconsin–Madison

Benson C, Abichou T, Albright W, Gee G, Roesler A (2001) Field evaluation of alternative earthen final covers, Int J Phytoremediation 3(1):1–21

Benson C, Bohnhoff G, Apiwantragoon P, Ogorzalek A, Shackelford C, Albright W (2004) Comparison of model predictions and field data for an ET cover. In: Tailings and Mine Waste '04. Balkema, Leiden, Netherlands:137–142

Benson C (2006) Numerical modeling in geoenvironmental practice, Geo-Strata–Geo Institute of ASCE 6(4):16–19

Bohnhoff G (2005) Water Balance Predictions and Field Data for Water Balance Covers in Semi-Arid Regions. MS Thesis, Geological Engineering, University of Wisconsin–Madison, Madison, Wisconsin, USA

Brooks R, Corey A (1966) Properties of porous media affecting fluid flow, J Irrig and Drainage 92(2):61–88

Fayer M (2000) UNSAT-H Version 3.0: Unsaturated Soil Water and Heat Flow Model – Theory, User Manual, and Examples. PNNL-13249. Pacific Northwest Laboratories, Richland, WA, USA

Gitirana G, Fredlund M, Fredlund D (2005) Infiltration-runoff boundary conditions in seepage analysis. In: Proc 58th Canadian Geotechnical Conf and 6th Joint IAH-CGS Conf, Saskatoon, SK, Canada

Meerdink J, Benson C, Khire M (1995) Unsaturated hydraulic conductivity of two compacted barrier soils, J Geotech Eng 122(7):565–576

Scanlon B, Christman M, Reedy R, Porro I, Šimunek J, Flerchinger G (2002) Intercode comparisons for simulating water balance of surficial sediments in semiarid regions, Water Res Res 38(12):1–16

Schaap M, Leij F (2000) Improved prediction of unsaturated hydraulic conductivity with the Mualem–van Genuchten model, Soil Sci Soc Am J 64:843–851

van Genuchten M (1980) A closed-form equation for predicting the hydraulic conductivity of unsaturated soils, Soil Sci Soc Am J 44:892–898

Mechanical Behaviour of a Clay Layer for Landfill Cap Cover Application: Experimental Investigation and Numerical Modelling

Sophie Camp, Jean-Pierre Gourc, Olivier Plé, and Pascal Villard

Lirigm, UJF Grenoble, BP 53, 38041 Grenoble cedex 9, France
sophie.camp@ujf-grenoble.fr, gourc@ujf-grenoble.fr,
olivier.ple@ujf-grenoble.fr, pascal.villard@ujf-grenoble.fr

Summary. Dedicated hazardous waste landfills have a cover barrier including, in particular, a compacted clay liner, which is a major element of the safety of the site. However, the implementation of this barrier poses many problems, in particular in terms of compaction. Moreover this cover must keep its sealing function during the lifetime of the landfill and the period of waste monitoring. However, potential differential settlements within the waste should be considered and this phenomenon can induce bending strains in the clay layer and create damage. Four point bending tests are performed in the laboratory to characterise the clay behaviour. Analysis is focused on the crack initiation and on the consequences of damage (crack propagation) on the mechanical behaviour. In addition a numerical simulation, by the discrete elements method, was carried out on the same type of tests. The comparison between the results from the numerical modelling and those from the experimental tests provides a method to determine the mechanical parameters of the material. These results are used to optimise the method of implementation of the clay on site and also to predict the mechanical behaviour of the material.

Key words: clay barrier, mechanical behaviour, numerical modelling

1 Introduction

Landfills have a top barrier including, in particular, a cap cover of compacted clay. This barrier must keep its watertightness properties during the life of the landfill (operation and after closure). However, this barrier meets many problems. In particular those related to its implementation and to the mechanical solicitations after closing the cell. Moreover, it is necessary to maintain the physical and mechanical characteristics of the clay barrier at a constant level during the life of the landfill. Currently, there is very little knowledge (and no specific regulations) concerning the behaviour of a fine soil under low confinement and subjected to differential settlements. As the clay material is very sensitive to flexibility (Jessberger and Stone 1991), four point bending tests are performed, in the laboratory, to characterise the tensile failure.

This paper describes a new experimental protocol for the bending tests and a new numerical technique of analysis. Comparisons between numerical and experimental results are presented. They contribute to a better understanding of the mechanical behaviour of the clay material. Finally, the practical aim is to make possible the selection of clay, in particular in terms of flexibility (to avoid cracking), and consequently to limit the thickness of the mineral barrier, whilst retaining an equivalent safety.

2 Improvement to the Experimental Protocol for the Bending test

2.1 Bending test Experimental Procedure

Experimental tests were carried out with a clay soil, mainly kaolinite ($< 2\,\mu\text{m}$: 5%; between $2\,\mu\text{m}$ and $80\,\mu\text{m}$: 67%; γd_{opt}: $17.9\,\text{kN/m}^3$; w_{opt}: 14.5%; PI: 13%) coming from a real site. Four point bending tests were selected and conducted in the LIRIGM laboratory (see Fig. 1). This kind of test has been selected for studying the influence of the bending moment in the central zone of the sample (Indraratna and Lasek 1996) and initiating the failure under tensile strength. The distance between the support rollers must be large enough to limit the influence of the compression zone. Bending tests were performed on samples with dimensions of 10 cm × 10 cm × 40 cm and carried out on samples compacted at different energies and different water contents. In this range of water content and of compaction energy the clay seems to be brittle because of the low value of the plasticity index.

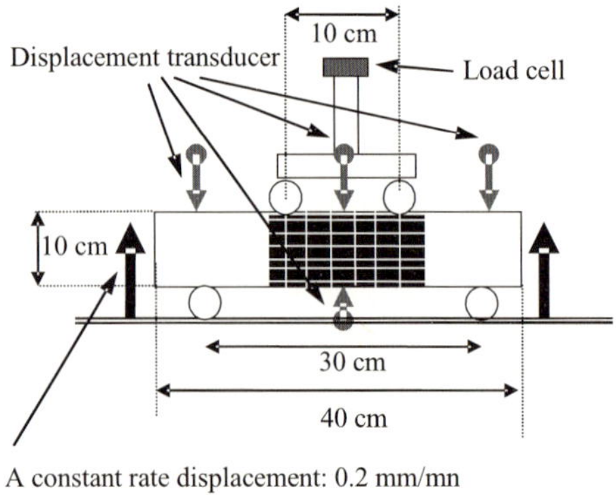

Fig. 1. Four point bending tests

Four displacement transducers are used, two are placed on the two support rollers, and two are placed in the middle of specimen (see Fig. 1). The load was measured by means of a load cell positioned on the upper axial support rollers of the press. A constant rate displacement (0.2 mm/min) was applied through the lower tray of the press. The middle part of the specimen, where the bending moment is constant and maximal, was painted in black and fitted with a series of white pins (see Fig. 2). This area was continuously recorded with a camera (5 million pixel with 1 pixel = 0.11 mm) to determine the initiation and propagation of cracks (see Fig. 2) and to estimate tensile strains by digital image correlation.

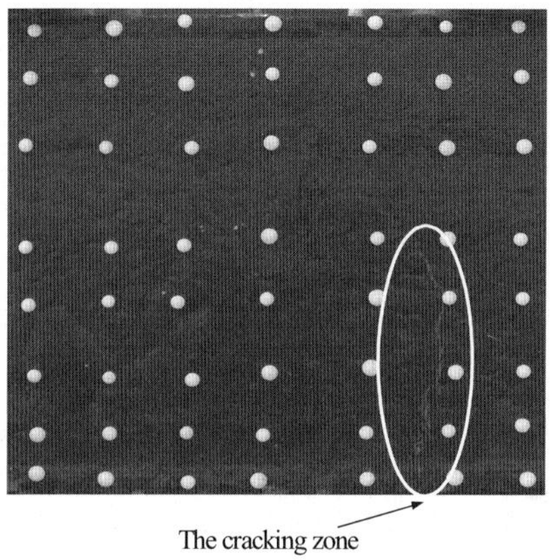

Fig. 2. Observation of the cracking zone

2.2 Bending Test Analysis and Results

Series of tests were carried out and our analysis focused on the cracking mode. The experimental results have been obtained with a good repeatability (dispersion of the maximum strength less than 15%) (see Fig. 3). The initial characteristics of the sample are fixed (water content: w = $16.5 \pm 0.8\%$ and dry density: $\gamma_d = 17.3 \pm 0.3\,\text{kN/m}^3$). No local punching is noted under the two upper rollers. The tensile cracking is localised in the lower part of the specimen (see Fig. 2). Qualitatively, a brittle behaviour is observed in the range of $55\,\text{kPa} < \sigma_{max} < 65\,\text{kPa}$ corresponding to F^* between 190 and 210 N (see Fig. 3) and for maximal tensile strains around $\varepsilon = 0.1\%$ (deducted from the image analysis). However, to quantify the mechanical behavior a DEM approach, with hypotheses, is necessary (Ammeri et al. 2006). This is the purpose of the following paragraph.

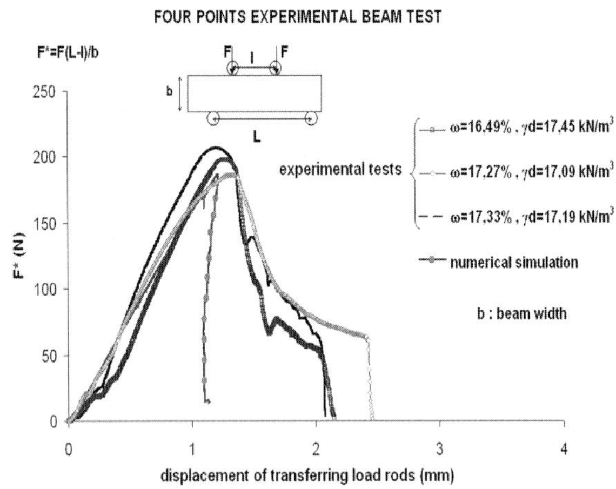

Fig. 3. Experimental and Numerical bending tests simulations

3 Discrete Element Approach Framework

3.1 Principle of the DEM Approach

The clay is considered as an assembly of particles with cohesions and frictions. The interest of this approach is to consider simple individual laws for contacts between particles. For the numerical software used (PFC^{2D}), these relationships relate the relative displacement with contact forces and rigidities under the form:

$$F_i^n(t) = K^n U_i^n(t) n_i\,, \quad \Delta F_i^s = -K^s \Delta U_i^s\,, \quad F_i^s(t) = F_i^s(t-1) + \Delta F_i^s \quad (1)$$

where F_i^n and F_i^s are respectively the normal and tangential forces in contact (i) between two particles with normal n_i, and U_i^n and ΔU_i^s the normal and relative tangential displacements in contact (i). Also (t) is a time step in the particle movement. The distinct element model is based on the Newton's second law of motion under the form:

$$F_i = m_i\left(\ddot{x}_i - g_i\right)\,, \quad M_i = I_i \dot{\omega}_i \quad (2)$$

where m_i is the particle's mass, $\ddot{x}_i$ is the acceleration, g_i is the volumetric gravity force of particle (i), M_i, I_i and $\dot{\omega}_i$ respectively are the resultant moment, the principal moment of inertia of the particle and the angular acceleration about the principal direction. The contacts are defined by five rheological parameters: normal and tangential stiffness (K^n and K^s), normal and shear cohesions ($C^n$and C^s) and friction coefficient (μ). The parameters of the two contacting particles are assumed to act in series (see Fig. 4). With the micro mechanic parameters the real contact conditions between all particles are

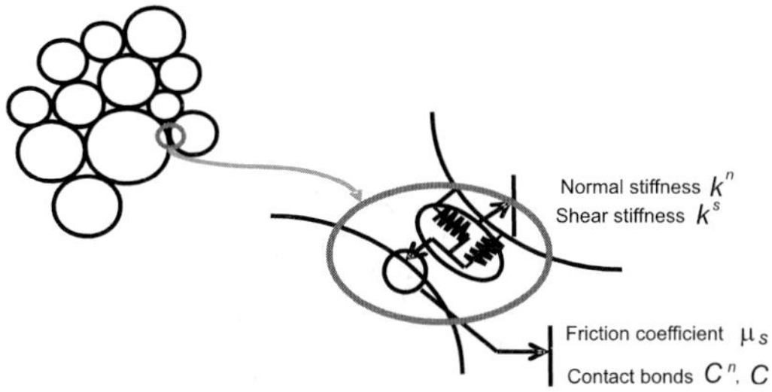

Fig. 4. The rheological model of the DEM method

fitted step by step. All reliable contacts are re-actualised according to the following conditions:

$$F^n > 0 \text{ (compression stress) } |F^s| \le F^s_{\max} \text{ with } F^s_{\max} = \max\{\mu F^n, C^s\} \quad (3)$$

$$F^n < 0 \text{ (tensile stress) } |F^n| \le C^n,\ F^s_{\max} = C^s. \quad (4)$$

The major difficulty in using this approach in real applications is the assignment of correct micro-properties to have the true relationship between micro and macro-properties. A preliminary parametrical numerical study is needed in order to fit the micro-parameters.

3.2 The Micro-Macro Relationship in the Simulation Approach

The fitting between the micro and macro parameters is made by carrying out direct compressive tests on the same clay material. Figure 5 shows the results of an experimental prediction of a compression test. The retained micro-mechanical properties of the Discrete Element Method are the normal and shear stiffness $K^n = K^s = 14\,\text{MN/m}$, the friction coefficient is equal to 0.8 and the normal and shear bonds $C^n = C^s = 300\,\text{kN/m}$.

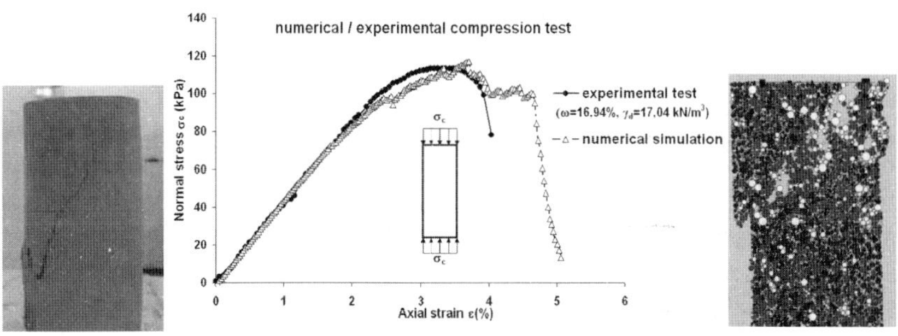

Fig. 5. An experimental prediction of unconfined compression test

3.3 Numerical Simulation of the Bending Test

To illustrate how the numerical approach has potential to predict the bending test results, a simulation of the experimental test shown in Fig. 3 was carried out, using micro parameters values calibrated from compression tests (Subsection 3.2). A good agreement with the experimental bending diagram is observed. Systematic comparison between numerical simulation and experimental results is given. The DEM approach provides a practical way to predict results for different bending beam tests. In particular, the brittle behaviour and the maximum stress before the initiation of crack ($\sigma_{\text{max}} = 45$ to $50\,\text{kPa}$ corresponding to $F^* = 150$ to $170\,\text{N}$) can be reproduced.

4 Conclusion

These tests were performed on a brittle clay. Clay samples with higher water contents (viscoplastic behaviour) were subjected to similar tests. This behaviour is more difficult to simulate by DEM and the further work is in progress.

The main target of this paper was to display the methodology used. The present study which shows the interest of using such an approach is integrated in the framework of a global project in the field of clay liner for hazardous waste.

The numerical study carried out with the PFC^{2D} code improves the potential of this method which can avoid multiplying experimental tests. The limitation of the elasticity method to predict the tensile strength was also improved. However, this method can be used to reproduced discontinuity (cracking propagation), the bending strength and the strains corresponding.

Acknowledgements

This work is supported by a CIFRE contract and in collaboration with Scetauroute, Andra and Benedetti. The authors wish to thanks these companies for providing the research team with the materials and laboratory test apparatus for this study.

References

Jessberger HL, Stone K (1991) Subsidence effect on clay barriers, Géotechnique 41:please give pages

Indraratna B, Lasek G (1996) Laboratory evaluation of the load-deflection behaviour of clay beams reinforced with galvanised wire netting, Geotextiles and geomembranes 14:555–573

Ammeri A, Jamei M, Guiras H, Bouassida M, Villard P, Plé O, Camp S, Gourc JP (2006) A Numerical study of compacted clay tensile strength by discrete element modelling: A bending test application. In: First Euro Mediterranean In Advance on Geomaterials and Structures, May 3–5, 2006 Hammamet, Tunisia

Finite Element Modelling of Soil-Vegetation Interaction

Buddhima Indraratna, Behzad Fatahi, and Hadi Khabbaz

School of Civil, Mining and Environmental Engineering, University of Wollongong, NSW 2522, Australia
indra@uow.edu.au, bf94@uow.edu.au, khabbaz@uow.edu.au

Summary. Behaviour of soils in the vadose zone is closely linked to water balance between ground and atmosphere. It seems that transpiration is the most uncertain and difficult to evaluate of all the terms in the soil water balance. The key variable to estimate the transpiration rate is the rate of root water uptake, which depends on the hydrological, geological and meteorological conditions. A mathematical model for the rate of root water uptake incorporating the root growth rate, ground conditions, type of vegetation and climatic parameters, has been developed. A conical shape is considered to represent the geometry of the tree root zone. Using this proposed model, the distribution of moisture and the matric suction profile adjacent to the tree are numerically analysed. Field measurements taken from literature published previously are compared with the authors' numerical model. The predicted results obtained from the numerical analysis, compared favourably with the field measurements, justifying the assumptions upon which the model was developed. The analysis also indicates that soil suction and settlement increase over the time, with the effect being more significant in the first stages of transpiration.

Key words: matric suction, settlement, root water uptake, transpiration, finite element, interaction

1 Introduction

Based on landslide investigation data and reports on control of structures in the vicinity of expansive and collapsible soils, most of recent failures were associated with hydrological and groundwater conditions less favorable than those assumed in designs. It seems that effects of hydrogeological features on soil properties are not very clear in geotechnical engineering designs. As Shen (1998) reported when the steep man-made slopes in Hong Kong are analysed using saturated shear strength parameters, it is often found that the factor of safety is less than one. In these cases, unsaturated soil suction is the main explanation. Indeed, the slopes will never be fully saturated even during severe rainstorm.

According to Nelson and Miller (1992), expansion and shrinkage are the result of changes in the soil water system that disturb the internal stress equilibrium. If the soil water chemistry alters either by changes to the volume of water or chemical composition, the inter-particle force field will change. If this resulting change in internal forces is not balanced by a corresponding change in the externally applied state of stress, the particle spacing will change to adjust the inter-particle forces until equilibrium is reached. This change in particle spacing manifests itself as shrinkage or swelling. Retarding evaporation, heavy rainfall and growth of trees and shrubs are the most important factors resulting in a noticeable change in the ground moisture. Trees, shrubs, and grasses deplete moisture from the soil through transpiration.

As transpiration is a continuous process of discharging water from the soil matrix, rainfall will be quickly taken in through tree roots. Thus, in a vegetated ground, effects of matric suction on shear strength and deformation of partially saturated soil must be considered. Moreover, Zhang et al. (2004) showed that rainfall does not necessarily eliminate the matric suction in the soil, and therefore, for a rigorous analysis it must be incorporated in design.

The main objective of this study is to establish a model for estimating root water uptake considering the growth rate of the plant, and then to develop an integrated transient model considering soil water extraction by roots within vadose zone to simulate the ground movement under the influence of vegetation. Developing an analytical solution to predict water flow in soil-vegetation porous media would be very complicated. Hence, numerical modelling becomes the choice to analyse and predict the movement of water. The results are then compared with field measurements to verify the numerical predictions. In addition, the effect of time on matric suction distribution and ground settlement is studied.

2 Root Water Uptake Model Description

Blight (1997) concluded that while agronomists, soil scientists and hydrologists have paid a great deal of attention to the interaction of ground and atmosphere, geotechnical engineers have tended to ignore it. Based on soil water balance theory, to compute water content of soil system, initial water content, the quantities of water entering and leaving the soil system need to be calculated. Thus, available water content of soil system can be estimated as

$$W = W_0 + I - O \tag{1}$$

where W is the available water content of the soil system, W_0 is the initial water content of the soil system, I is the quantity of water entering the soil system (Input), and O is the quantity of water leaving the soil system (Output). Considering the active root zone which is the zone vegetation roots develop in, and passive zone for soil profile, and including the hydrogeological aspects,

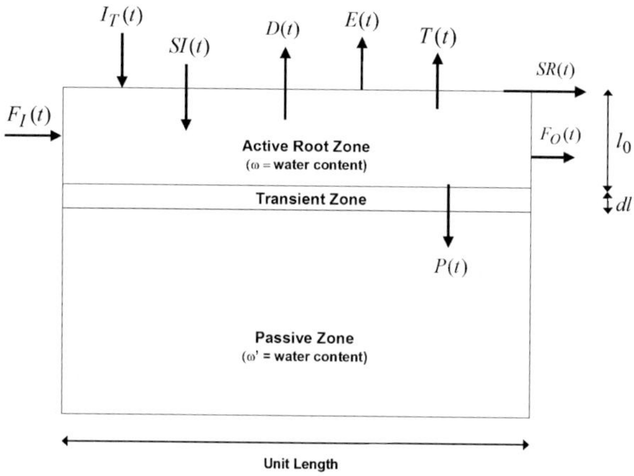

Fig. 1. Soil system including hyrdogeological features

water content of soil can be determined. Figure 1 illustrates the soil profile and the acting hydrogeological features. Based on mass conservation theory (Eq. (1)), following equations can be obtained

$$\omega(t)l(t) = \omega_0 l_0 + \int_0^t Z(t)\,\mathrm{d}t + \int \omega'(t)\frac{\partial l}{\partial t}\,\mathrm{d}t \tag{2}$$

$$\begin{aligned} Z(t) = {} & I_T(t) + SI(t) + F_I(t) - E(t) \\ & - T(t) - D(t) - P(t) - F_O(t) - SR(t) - \Delta S(t) \end{aligned} \tag{3}$$

where ω_0 is the initial water content of active root zone, $l(t)$ active root zone depth at time t, $I_T(t)$ is the effective interception, which reaches the soil system (precipitation below the tree canopy), $SI(t)$ is the supplemental irrigation rate in the soil system, $F_I(t)$ is the inflow of groundwater (lateral flow), $E(t)$ is evaporation rate, $T(t)$ is transpiration rate, $D(t)$ is drainage rate, $P(t)$ is percolation rate, $F_O(t)$ is the outflow of groundwater (lateral flow), $SR(t)$ is the surface runoff, which is equal to $SR(t) = SR_O(t) - SR_I(t)$, $SR_I(t)$ is the input surficial flow to soil system, $SR_O(t)$ is the output surficial flow from soil system, and $\Delta S(t)$ is the increase rate in surface storage. Blight (2003) drew the conclusion that the evapotranspiration term is probably the most uncertain and difficult to evaluate of all the terms in the soil water balance. According to Raudkivi (1979), the amount of evaporation can be estimated properly using different methods including evaporation pan, empirical formulae, water budget method, mass transfer method, or energy budget method. Consequently, the transpiration is the most uncertain part in estimation of eavapotraspiration which affects distribution of soil moisture content. Blight (2003) investigation pleaded that more in depth research and measurements are required to understand the phenomenon of transpiration by trees for proper prediction of the lateral and depth extents of the drying zones of various tree species.

Previous research by Indraratna et al. (2006) attempted to develop a mathematical model for the distribution of tree root water uptake within the root zone. This proposed model attempts to combine the effects of soil matric suction, root density and potential transpiration rate. Accordingly, a mathematical model is formulated to represent the rate of tree root water uptake as

$$S(x, y, z, t) = f(\psi)G(\beta)F(T_P) \tag{4}$$

where $G(\beta)$ is the root density factor, $f(\psi)$ is the soil suction factor, and $F(T_P)$ is the potential transpiration factor. As discussed by Indraratna et al. (2006), an appropriate representation for $f(\psi)$ based on Feddes et al. (1978) may be considered as follows

$$f(\psi) = \begin{cases} 0, & \text{if } \psi < \psi_{\text{an}}, \\ 1, & \text{if } \psi_{\text{an}} \leq \psi < \psi_d, \\ \dfrac{\psi_w - \psi}{\psi_w - \psi_d}, & \text{if } \psi_d \leq \psi < \psi_w, \\ 0, & \text{if } \psi_w \leq \psi, \end{cases} \tag{5}$$

where ψ_w is the soil suction at wilting point, the limit at which a particular vegetation is unable to draw moisture from the soil; ψ_d and ψ_{an} (soil suction at anaerobiosis point) are the highest and the lowest values of ψ at $S = S_{\text{max}}$, respectively, while S_{max} denotes the maximum rate of root water uptake.

Equation (6) for the root density factor and Eq. (7) for potential transpiration factor have been suggested by the authors

$$G(\beta) = \frac{\tanh\left(k_3\beta_{\text{max}} \exp(-k_1|z - z_0| - k_2|r - r_0|)\right)}{\int_{V(t)} \tanh\left(k_3\beta_{\text{max}} \exp(-k_1|z - z_0| - k_2|r - r_0|)\right) \mathrm{d}V} \tag{6}$$

$$F(T_P) = \frac{T_P(1 + k_4 z_{\text{max}} - k_4 z)}{\int_{V(t)} G(\beta)(1 + k_4 z_{\text{max}} - k_4 z)\,\mathrm{d}V} \tag{7}$$

where k_1 and k_2 are two empirical coefficients depending on the tree root system and type, k_3 is an experimental coefficient representing the influence of root density, z is the vertical coordinate (downward is positive), r is the radial coordinate, β_{max} is the maximum density of root length located at the point $(r, z) = (r_0, z_0)$, T_p is the rate of potential transpiration, k_4 is an experimental coefficient to involve depth on the potential transpiration distribution, and $V(t)$ is the volume of the root zone at time t. Discussion about the influence of above parameters on the rate of root water uptake (Eq. (4)) can be found in Fatahi and Indraratna (2006).

Borg and Grimes (1986) reviewed the root growth data of 48 crop species and found that increase of root zone dimensions with time delineates a sigmoidal curve which is a single sine function. Therefore,

$$z_{\text{max}}(t) = z_{\text{max}\,f} \cdot \{0.50 + 0.5 \sin\{3.03(t/t_f) - 1.47\}\}, \tag{8}$$

$$r_{\text{max}}(t) = r_{\text{max}\,f} \cdot \{0.50 + 0.5 \sin\{3.03(t/t_f) - 1.47\}\} \tag{9}$$

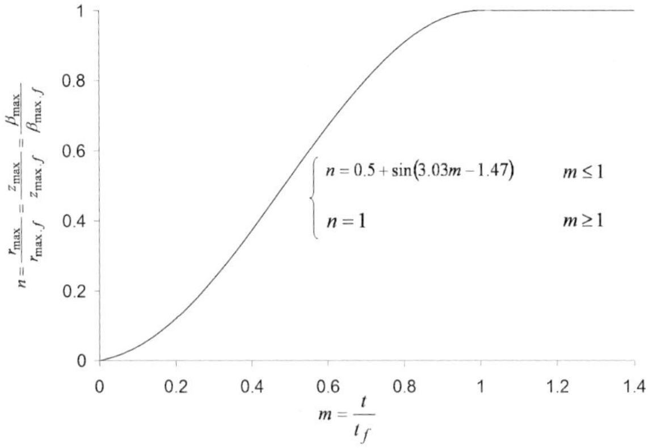

Fig. 2. Variation of normalized root zone size and maximum root density with normalised time

where $z_{\max}(t)$ is the maximum depth of root zone at time t, $r_{\max}(t)$ is the maximum lateral distance of root zone at time t, $z_{\max.f}$ is the maximum possible root zone depth, $r_{\max.f}$ is the maximum possible lateral distance of root zone and t_f is the time that tree's growth stops and root zone reaches to its maximum. Figure 2 illustrates the variation of normalized root zone size with normalised time.

According to Gerwitz and Page (1974), the effect of time on the maximum root length density are described by the same sigmoidal curve:

$$\beta_{\max}(t) = \beta_{f_{\max}} \cdot \{0.5 + 0.5 \sin[3.03(t/t_f) - 1.47]\} \tag{10}$$

Combining the exponential root length density suggested by Indraratna et al. (2006) and Eq. (10) yields

$$k_1(t) = \frac{k_{1f}}{0.5 + 0.5 \sin[3.03(t/t_f) - 1.47]} \tag{11}$$

$$k_2(t) = \frac{k_{2f}}{0.5 + 0.5 \sin[3.03(t/t_f) - 1.47]} \tag{12}$$

As a result, according to Eq. (4), the rate of root water uptake $S(x, y, z, t)$ can be estimated by multiplying three functions that represent the soil suction effect $f(\psi)$, the root density effect $G(\beta)$, and the potential transpiration effect $F(T_P)$. To calculate $f(\psi)$, Eq. (5) suggested by Feddes et al. (1978) to determine the effects of soil suction, is used in this study. In order to determine $G(\beta)$, Eq. (6) incorporating Eqs. (8)–(12) are employed, and to estimate $F(T_P)$, Eq. (7) in conjunction with Eqs. (8) and (9) is adopted.

3 Finite Element Simulation

To verify the model, a case history reported by Biddle (1998) has been considered for a single, 20 m high Poplar tree in Cambridge (U.K.). The tree is located in an area of mown grass on Gault clay, and forms part of a double row of Poplars. As reported by Biddle (1998), all soil moisture measurements were conducted using a neutron soil moisture gauge. The estimated parameters based on the available literature are given in Table 1. Figure 3 illustrates the mesh and element geometry and boundary conditions of the finite element model built in ABAQUS code. A two-dimensional plane strain mesh employing 4-node bilinear displacement and pore pressure elements (CPE4P) was considered. Because of symmetry, a zero flux boundary was applied along the left boundary. The flux boundary at the surface is controlled by both climatic conditions and soil properties. In this study, it is assumed that rainfall and evaporation are in balance and thus a "no water in-flow" condition is applied at the surface. The developed theoretical model representing the rate of root water uptake distribution was included in the FE analysis through appropriate Fortran subroutines. In addition, it is assumed that the tree is well grown.

This numerical analysis is based on the effective stress theory of unsaturated soils incorporated in the ABAQUS finite element code. The effective stress in the unsaturated soil is given by Bishop (1959):

$$\sigma'_{ij} = \sigma_{ij} - u_a\delta_{ij} + \chi(u_a - u_w)\delta_{ij} \tag{13}$$

where σ'_{ij} is the effective stress of a point on a solid skeleton, σ_{ij} is the total stress in the porous medium at the point, u_a is the pore air pressure, u_w is the pore water pressure, δ_{ij} is Kronecker's delta ($\delta_{ij} = 1$ when $i = j$ and $\delta_{ij} = 0$ when $i \neq j$), and χ is the effective stress parameter attaining a value of unity for saturated soils and zero for dry soils. Fredlund and Morgenstern (1977) proposed a constitutive model based on the concept of independent stress state variables, in which the total stress tensor, the water and air pressures are considered as independent features. The state of stress model requires a

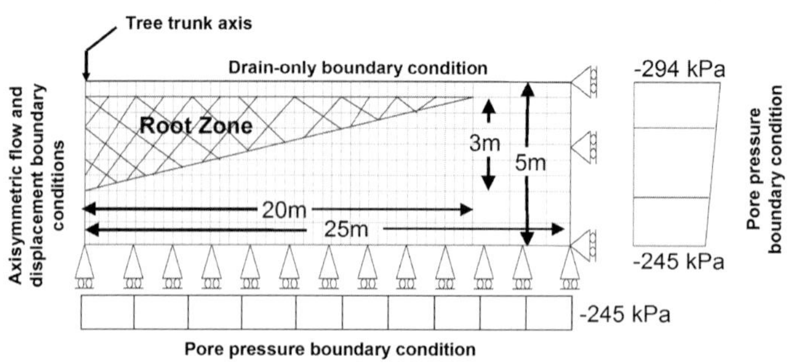

Fig. 3. The geometry and boundary conditions of the FE model (after Fatahi and Indraratna (2006))

Table 1. Parameters applied in the finite element analysis

Parameter	Value	Reference/Comments
ψ_{an}	4.9 kPa	Feddes et al. (1978); Clayey soil with air content of 0.04
ψ_w	1500 kPa	Feddes et al. (1978); $1500 < \psi_w < 2000$ kPa
ψ_d	40 kPa	Feddes et al. (1978); $40 < \psi_d < 80$ kPa
γ	20 kN/m^3	Samuels (1975); Typical unit weight for Gault clay
$r_{max.f}$	20 m	Biddle (1998); Estimated from field measurements ($17\,\text{m} < r_{max} < 23\,\text{m}$)
$z_{max.f}$	1.5 m	Biddle (1998); Estimated from field measurements
$(k_s)_h$	5×10^{-9} m/s	Terzaghi et al. (1996); Typical saturated permeability of unfissured and unweathered clay
$(k_s)_v/(k_s)_h$	1/10	Smith and Smith (2004); $1/10 \leq (k_v)_s/(k_h)_s \leq 1/5$
PI	41	Biddle (1998); Measured plasticity index
e_0	1.25	Samuels (1975); Typical void ratio of Gault clay
C_s	0.023	Ng (1998); Swelling index of heavily over consolidated clay
(r_0, z_0)	(3m, 1m)	Radial and vertical coordinate of the maximum root density point
$\beta_{f.max}$	25 m^{-2}	Taken from the general shape suggested by Landsberg (1999)
k_3	0.0874 m^{-1}	As above
k_4	0.014	Coefficient of potential transpiration distribution
$k_1.f$	5	Coefficient of vertical root distribution
$k_2.f$	0.50	Coefficient of horizontal root distribution
ν	0.30	Almeida et al. (1986); Typical Poisson's ratio of Gault clay
T_P	45 l/day	Schneider et al. (2002)
Passing #200	95%	Typical value for Gault clay (5% fine sand)

considerable amount of laboratory testing to identify the model parameters which is time consuming and cost prohibitive and also different plasticity models for saturated and unsaturated soils need to be used in the state of tress theory. Consequently, in this study with its own complexities and numerous measured and assumed parameters, the effective stress theory (Eq. (13)) is being used. Bishop's effective stress concept for predicting shear strength and volume change in unsaturated soils has recently been discussed and validated by Khalili et al. (2004). In addition, Khabbaz (1997) presented a relationship for χ as a function of matric suction and the air entry value. The soil-water characteristic curve employed in this study is shown in Fig. 4.

The finite element analysis is conducted in two stages: (i) geostatic and (ii) consolidation. The first stage is to ensure that the analysis commences from a state of equilibrium under geostatic loading. The consolidation stage is to avoid non-physical oscillations and possible divergence problems caused by

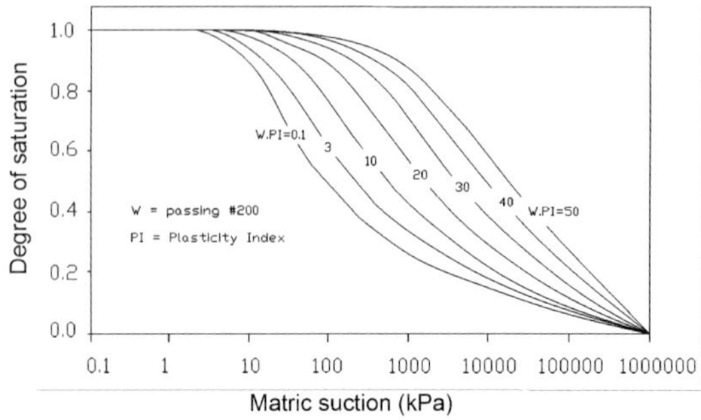

Fig. 4. Predicted soil water characteristic curve (after Zapata et al. (2000))

non-linearities. This stage includes a transient analysis of partially saturated soil under transpiration, starting with 1-day intervals and then continued for five months from the middle of spring until the end of September.

The coefficient of unsaturated soil permeability has been calculated based on Brooks and Corey (1964), thus:

$$k = k_s(e)S_e^{\frac{2+3\lambda}{\lambda}} \tag{14}$$

where $k_s(e)$ is the saturated coefficient of permeability estimated based on the well known Kozeney–Carman equation, λ ($= \Delta \log S_e / \Delta \log \psi$) is the slope of the soil water characteristic curve on a log-log plot, and S_e is the effective degree of saturation.

As fluid passes through a porous medium, a coupled flow-deformation analysis of unsaturated soil is required to capture the 3-phase interaction among the soil, air, and water. The governing equations for pore fluid diffusion and deformation are a combination of Eq. (1) and the relevant deformation equations. The over-consolidated behaviour of Gault clay is defined by the following consolidation equation:

$$\mathrm{d}e^{\mathrm{el}} = C_s \ln\left(\frac{p_0 + \mathrm{d}p}{\mathrm{d}p}\right) \tag{15}$$

where $\mathrm{d}e^{\mathrm{el}}$ denotes the change of void ratio in the element, C_s is the swelling index, p_0 is the initial mean effective stress, and $\mathrm{d}p$ is the change in mean effective stress on the soil skeleton.

The deformation in the soil profile due to the root water uptake is predicted through a coupled flow-deformation analysis, considering stress-deformation equations. The ground settlements at various depths after five months of continuous transpiration are shown in Fig. 5. Base on Fig. 5, the ground settlement decreases rapidly with distance up to 12 m. The settlement also decreased rapidly with depth. At ground surface, 8.8 mm vertical settlement at the tree location increased to 9.2 mm at 3 m distance from the tree trunk and then decreased sharply to 2 mm at 13.5 m away from the tree trunk.

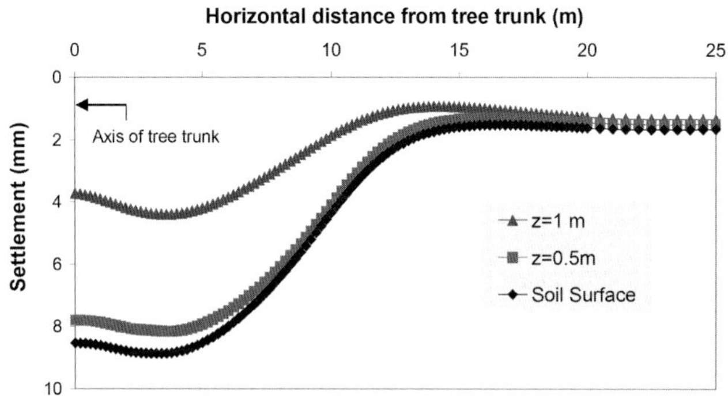

Fig. 5. Ground surface settlement profile at various depths

Figure 6 shows a comparison between the field measurements and numerical predictions for a reduction in moisture content. Numerical analysis predictions based on the authors' model for root water uptake are in acceptable agreement with the field measurements reported by Biddle (1998). It is important to note that in the numerical analysis, root water uptake as a sink term were considered in the flow equation, but the effect of each root was not considered individually. As the main roots penetrate into the soil, there may be a gap between them, which can lead to water collecting in the gap. Since the woody roots are in a denser pattern under the trunk and in close proximity, a disparity between the field measurements and predictions in this area seems more likely. Furthermore, the actual field data is probably influenced by the soil heterogeneity.

4 Effect of Time (t)

The soil suction variation and soil consolidation are time dependent variables. Therefore, investigating the effect of elapsed time on the soil suction profile

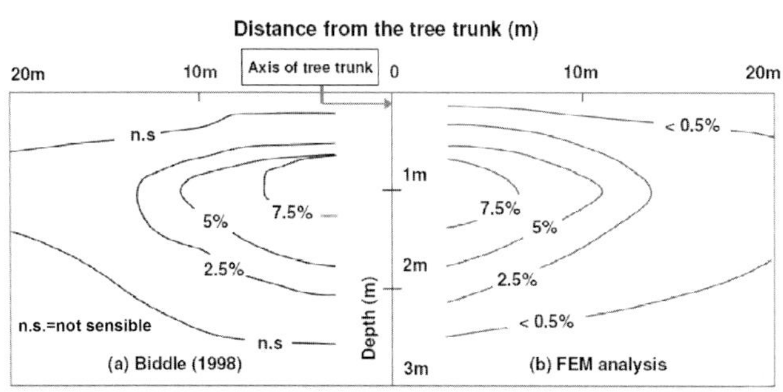

Fig. 6. Contours of moisture content reduction in the vicinity of a poplar tree (**a**) Biddle (1998) (**b**) FEM analysis (after Fatahi and Indraratna (2006))

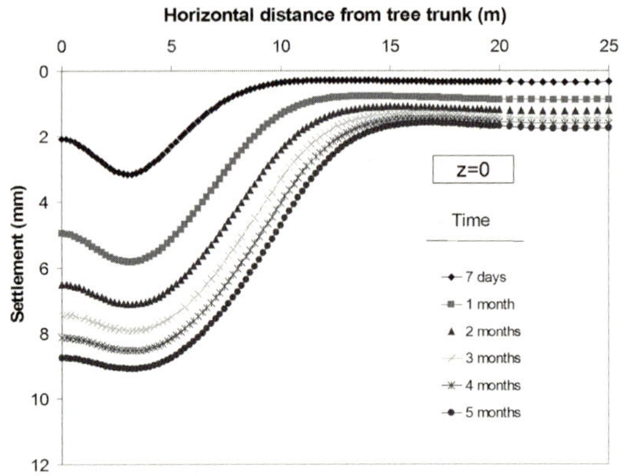

Fig. 7. Effect of elapsed time on soil surface settlement

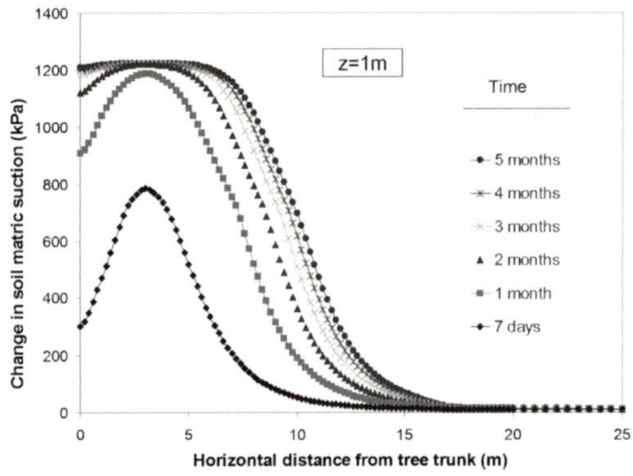

Fig. 8. Matric suction change at 2 m depth

and the ground settlement are required for the design purposes. To study the effect of elapsed time, the results of the reference case study are compared with five different time intervals, from seven days to five months as shown in Figs. 7 and 8.

Figure 7 indicates that the ground settlement increases rapidly with time. The maximum ground surface settlement of 3 mm after seven days increases to 9.2 mm in five months. Figure 8 presents the change of matric suction along the horizontal distance from the tree trunk associated at 1 m depth.

As expected and shown in Figure 8, the generated matric suction induced by tree transpiration increases by the time. Figure 8 shows that the peak matric suction increases about 50% when elapsed time increases from seven days to two months. According to Fig. 8, after two months of continuous transpiration, the wilting point suction is reached at point (r_0, z_0) and then develops around this point.

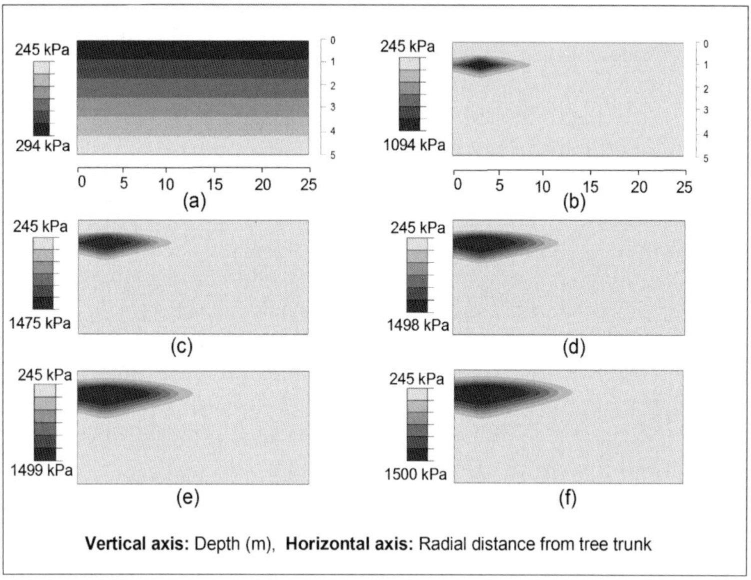

Fig. 9. Evolution of soil matric suction (kPa), (**a**) initial (**b**) after seven days, (**c**) one month, (**d**) two months, (**e**) four months, and (**f**) six months

Figure 9 shows the evolution of soil matric suction at different times. Knowing the moisture content variation, the development of matric suction can be predicted reasonably well using the soil water characteristic curve. As Fig. 9 indicates, the matric suction established in the root zone propagates radially and influences the surrounding soil.

5 Conclusions

A numerical finite element model was developed to examine the effects of tree transpiration on ground condition. The developed model takes into account the coupled flow-deformation equations. The finite element discretisation was formulated using partially saturated soil elements, which are capable of capturing the role of unsaturated permeability and the degree of saturation at various levels of matric suction. Tree root suction was included in the analyses through the model developed by the authors which takes into account soil matric suction, and distribution of root density, potential transpiration and root growth rate.

Comparing the numerical results and field measurements of the moisture content in the vicinity of a Poplar tree indicated that the numerical analysis incorporating the proposed model could predict the variation of moisture content surrounding the tree. As transpiration and soil consolidation are time dependent functions, influence of elapsed time has been investigated by conducting a transient ground analysis. Results indicate that soil matric suction and vertical deformation increase over the time, with the effect being more important in the beginning of transpiration.

The developed model may improve the design and construction procedures for railway lines, foundations, slopes, and embankments near native vegetation. As the influence zone of each tree can be several meters in radius, the methodological planting of native trees along rail corridors at a practical distance away from the track is currently considered by rail organisations. Native vegetation generating soil suction is comparable to the role of prefabricated vertical drains with vacuum pressure, in terms of improved drainage, and associated increase in shear strength. In addition, the tree roots provide a natural reinforcement effect, which the current model has not simulated thus far. Considering various soil properties, the type of vegetation and atmospheric conditions, the proposed mathematical model would be most useful to predict the formation behaviour in a rail track environment.

Acknowledgement

This research has been sponsored by the Australian Cooperative Research Centre for Railway Engineering and Technologies (Rail-CRC). The contributions and feedback from various industry colleagues, particularly Wayne Potter and David Christie, are appreciated. The assistance of Dr. Don Cameron (University of South Australia) is also acknowledged.

References

Almeida MSS, Britto AM, Parry RHG (1986) Numerical modelling of a centrifuged embankment on soft clay, Canadian Geotechnical Journal 23(2):103–114

Biddle PG (1998) Tree Root Damage to Buildings. Willowmead Publishing Ltd., Wantage

Bishop AW (1959) The principle of effective stress, Teknish Ukebadk 106(39):859–863

Blight GE (1997) Interactions between the atmosphere and the earth, Geotechnique 42(4):715–766

Blight GE (2003) The vadose-zone soil-water balance and transpiration rates of vegetation, Geotechnique 53(1):55–64

Borg H, Grimes DW (1986) Depth Development of Roots with Time: An Empirical Description, Trans ASAE 29(1):194–197

Brooks RH, Corey AT (1964) Hydraulic properties of porous media, Colorado, U.S.A., Colorado State University, Report No. 3

Fatahi B., Indraratna B (2006) A Case Study and Pilot Parametric Study on the Effect of Root-Based Suction on Ground Behaviour, Proceedings of the Institution of Civil Engineers, Geotechnical Engineering (submitted)

Feddes RA, Kowalik PJ, Zaradny H (1978) Simulation of field water use and crop yield, Simulation Monograph, Pudoc, Wageningen

Fredlund DG, Morgenstern NR (1977) Stress State Variables for Unsaturated Soils, ASCE J Geotech Div 103(5):447–466

Gerwitz A, Page ER (1974) An empirical mathematical model to describe plant root systems, Journal of Applied Ecology 11:773–782

Indraratna B, Fatahi B, Khabbaz H (2006) Numerical analysis of matric suction effects of tree roots, Proceedings of the Institution of Civil Engineers, Geotechnical Engineering 159(2):77–90

Khabbaz H (1997) An effective stress theory based numerical model for flow and deformation in unsaturated soils. PhD Thesis, University of NSW, Australia

Khalili N, Geiser F, Blight GE (2004) Effective stress in unsaturated soils: review with new evidence, International Journal of Geomechanics 4(2):115–126

Landsberg JJ (1999) The Way Trees Use Water, Water and Salinity Issues in Agroforestry No. 5, RIRDC Publication No. 99/37, Australia 1–24

Nelson JD, Miller DJ (1992) Expansive Soils, Problems and Practice in Foundation and Pavement Engineering, John Wiley & Sons, New York

Ng CWW (1998) Observed performance of multopropped excavation in stiff clay, Journal of Geotechnical and Geoenvironmental Engineering 124(9):889–906

Raudkivi AJ (1979) Hydrology. Pergamon Press

Samuels SG (1975) Some properties of the Gault Clay from Ely-Ouse Essex water tunnel, Geotechnique 25(2):239–264

Schneider WH, Hairsh SR, Compton HR, Burgess AE, Wrobel JG (2002) Analysis of hydrologic data to evaluate phytoremediation system performance, Proceedings of The Third International Conference on Remediation of Chlorinated and Recalcitrant Compoind, Battelle Press, Columbus, Ohio

Shen JM (1998) Soil Suction in relation to slope stability: A summary of research carried out in Honh Kong in 1978–1997, Proceeding of the annual seminar on slope engineering in Hong Kong 93–99

Smith GN, Smith IGN (2004) Elements of Soil Mechanics. Blackwell Publishing, Oxford

Standards Australia (1996) Residential slabs and footings construction, AS2870–1996

Terzaghi K, Peck RB, Mesri G (1996) Soil Mechanics in Engineering Practice. John Wiley & Sons, Inc., U.S.A.

Uniform Building Code (1997) International Building Code. International Code Council, U.S.A.

Zapata CE, Houston WN, Houston SL, Walsh KD (2000) Soil-water characteristic curve variability. In: Shackeford et al. (eds) Advances in unsaturated geotechnics. GEO-Institute, ASCE, U.S.A. 84–124

Zhang LL, Fredlund DG, Zhang LM, Tang WH (2004) Numerical study of soil conditions under which matric suction can be maintained, Canadian Geotechnical Journal 41(4):569–582

Contaminant Transport Modelling Using EFGM

Praveen Kumar Rachakonda, Gouda Ramanagouda Dodagoudar, and Nageswara Rao Bhairavavajjula

Department of Civil Engineering, Indian Institute of Technology Madras, Chennai – 600 036, India
praveenkumar_r@smail.iitm.ac.in, goudar@iitm.ac.in, bnrao@iitm.ac.in

Summary. Modelling of contaminant transport in unsaturated soils is necessary to achieve an improved understanding of pollutant migration in the vadose zone, which differs significantly from that of fully saturated porous media. This paper deals with the application of a meshless method called Element Free Galerkin Method for modelling the contaminant transport through the unsaturated porous media with transient flow condition. A one dimensional advection-dispersion equation with linear retardation is considered. The unknown concentration values are approximated by using moving least square approximants, which are constructed by using a weight function, a linear basis function and a set of non-constant coefficients. The weak form of the governing equation is formulated and Lagrange multiplier method is used for enforcing the essential boundary condition. The van Genuchten model is used for describing the hydraulic properties of the soil. A numerical example is presented to illustrate the applicability of the proposed method and the results are compared with those obtained from the finite element method.

Key words: contaminant transport, element free Galerkin method, continuous source, partially saturated soil, van Genuchten model

1 Introduction

Contaminants dispersed at the land surface migrate through the unsaturated zone before reaching the saturated zone. The behavior and fate of contaminant in the vadose zone is complex as it is controlled by the interactions occurring between dispersion, advection, sorption and degradation processes. Many researchers proposed various numerical models, such as Finite Difference Method, Finite Element Method and Boundary Element Method, which incorporate the above processes, for solving the governing equation of the contaminant transport model. Obtaining numerical solutions for the contaminant transport equation is difficult, because they require special methods with artificial viscosity, upwinding, etc. for solving hyperbolic type equations i.e. for advection dominant problems and also accuracy of the solution depends on the grid/mesh that is generated for solving the equation.

In the past decade, new numerical techniques called meshless methods, i.e. Element Free Galerkin Method (EFGM), Smooth Particle Hydrodynamics, Reproducing Kernel Particle Method, etc. have been developed. In these methods the approximate solutions are constructed entirely in terms of set of nodes and no elements or characterization of the interrelationship of the nodes are needed to construct the discrete equations. Among all the meshless methods, EFGM developed by Belytschko et al. (1994) has been successfully used for solving numerous problems related to various fields of study.

In this paper, the EFGM is used to solve the one dimensional advection dispersion contaminant equation with linear retardation term, for the unsaturated porous media with transient flow condition. The van Genuchten model (1980) is used to describe the hydraulic properties of the soil. Student's t distribution function is used as a weight function in the meshless analysis. As the EFGM does not satisfy the Kronecker delta function, Lagrange multiplier method is used to enforce the essential (Dirichlet) boundary condition. A `MATLAB` program is developed to code the procedure of the EFGM for contaminant migration. The results obtained are compared with that of Wamos – T model developed by Diaw et al. (2001).

2 Element Free Galerkin Method

The discretization of the governing equations by EFGM requires Moving Least Squares interpolation function (Lancaster and Salkauskas 1981) which is made up of three components: a weight function associated with each node, a monomial basis function and a set of non-constant coefficients.

2.1 Moving Least Square (MLS) Approximations

Consider a continuous function C defined on a domain Ω. To approximate the distribution of function C in Ω, the global approximation form $C^h(x)$ is defined as follows (Belytschko et al. 1994):

$$C(x) \cong C_L^h(x) = \sum_{i=1}^{m} p_i(x) a_i(x) = \mathbf{p}^T(x)\mathbf{a}(x) \quad \forall \quad x \in \Omega \tag{1}$$

where

$$\begin{aligned} \mathbf{p}^T(x) &= [p_1(x)\ p_2(x)\ \dots\ p_m(x)], \\ \mathbf{a}^T(x) &= [a_0(x)\ a_1(x)\ a_2(x)\ \dots\ a_m(x)] . \end{aligned} \tag{2}$$

$\mathbf{p}(x)$ is a basis function and $\mathbf{a}(x)$ is a vector of undetermined coefficients, whose value can vary according to the position of x in Ω and m is the order of basis. The coefficient vector $\mathbf{a}(x)$ at any point is determined by minimizing L_2 error norm of weighted least squares. The L_2 error norm is defined as follows:

$$J = \sum_{I=1}^{n} w(x - x_I)[C_L^h(x_{I,x}) - C_I]^2 = \sum_{I=1}^{n} w(x - x_I)[\mathbf{p}^T(x_I)\mathbf{a}(x) - C_I]^2 \quad (3)$$

where $w(x - x_I)$ is the weight function associated with the nodal point x_I in the domain Ω and n is a local node number whose support includes x. By minimizing J with respect to $a(x)$ gives

$$\mathbf{A}(x)\mathbf{a}(x) = \mathbf{B}(x)\mathbf{C} \Rightarrow \mathbf{a}(x) = \mathbf{A}^{-1}(x)\mathbf{B}(x)\mathbf{C} \quad (4)$$

where

$$\begin{aligned} \mathbf{A}(x) &= \sum_{I=1}^{n} w(x - x_I)\mathbf{p}(x_I)\mathbf{p}^T(x_I)\,, \\ \mathbf{B}(x) &= [w(x - x_1)\mathbf{p}(x_1) \;\ldots\; w(x - x_n)\mathbf{p}(x_n)]\,, \\ \mathbf{C}^T &= [C_1, C_2, \ldots, C_n] \end{aligned} \quad (5)$$

Therefore

$$\begin{aligned} C(x) \cong C^h(x) &= \sum_{I=1}^{n} \Phi_I(x)C_I = \sum_{I=1}^{n} \mathbf{p}^T\mathbf{A}^{-1}\mathbf{B}_I C_I = \mathbf{\Phi}\mathbf{C}\,, \\ \Phi_I(x) &= \mathbf{p}^T\mathbf{A}^{-1}\mathbf{B}_I, \qquad \mathbf{\Phi} = \{\Phi_1\ \Phi_2\ \Phi_3\ \cdots\ \Phi_n\} \end{aligned} \quad (6)$$

where Φ_I is a MLS shape function of node I.

The spatial derivatives of the MLS shape function of node I can be obtained by

$$\Phi_{I,x} = (\mathbf{p}^T\mathbf{A}^{-1}\mathbf{B}_I)_{,x}\mathbf{p}^T_{,x}\mathbf{A}^{-1}\mathbf{B}_I + \mathbf{p}^T(\mathbf{A}^{-1})_{,x}\mathbf{B}_I + \mathbf{p}^T\mathbf{A}^{-1}\mathbf{B}_{I,x}\,. \quad (7)$$

2.2 Weight Function Description

Weight function is non zero only over a small neighbourhood ofx_I, called the domain of influence of node I. In this paper student's t distribution weight function (Rao and Rahman 2000) is adopted, which is given below. The weight function is written in terms of normalized radius r as

$$w(x - x_I) \equiv w(r) = \begin{cases} \dfrac{(1+\beta^2 r^2)^{-(1+\beta)/2} - (1+\beta^2)^{-(1+\beta)/2}}{1 - (1+\beta^2)^{-(1+\beta)/2}} & r \le 1\,, \\ 0 & r > 1\,, \end{cases} \quad (8)$$

where $r = d_I/d_{mI}$, $d_I = \|x - x_I\|$, $d_{mI} = d_{\max} z_I$. z_I is the distance to the nearest node in neighbourhood and $d_{\max}$ is a scaling factor.

3 Discritization of Governing Equation

The one dimensional form of the governing equation for contaminant transport through the unsaturated porous media is:

$$\frac{\partial}{\partial t}(R\theta C) = \frac{\partial}{\partial x}\left(\theta D\frac{\partial C}{\partial x}\right) - \frac{\partial}{\partial x}(uC), \quad R = 1 + \frac{\rho_d K_d}{\theta} \tag{9}$$

Initial Condition at $t = 0$: $C = C_i$ in Ω
Boundary Conditions

$$C(0,t) = C_0, \text{ in } \Gamma_s \text{ (Dirichlet Boundary Condition)} \tag{10a}$$

$$\frac{\partial C}{\partial x} n_s = g, \text{ in } \Gamma_E \text{ (Neumann Boundary Condition)} \tag{10b}$$

where θ is the volumetric water content of soil, ρ_d is the bulk density of soil, K_d is the distribution coefficient, C is the concentration of contaminant, D is the dispersion coefficient, u is the seepage velocity, R is the retardation coefficient, C_0 and g are the concentration of contaminant at the source and concentration gradient at the exit boundary respectively, n_s is a unit normal to domain Ω and Γ_s and Γ_E are the portions of boundary Γ where source concentration and concentration gradient are prescribed.

The hydrodynamic properties of the soil are described by the functions of van Genuchten model (1980):

$$S = \begin{cases} \dfrac{1}{\left[1 + (\alpha|h|)^{\chi}\right]^{1-\frac{1}{\chi}}} & \text{if } h \le 0\,, \\ 1 & \text{if } h \ge 0 \end{cases} \tag{11a}$$

and

$$K = K_s(S)^{0.5}\left[1 - \left(1 - S^{\frac{\chi}{\chi-1}}\right)^{1-\frac{1}{\chi}}\right]^2 \quad \text{for} \quad \chi > 1 \tag{11b}$$

where θ_r and θ_s are the residual and saturated volumetric water contents of the soil respectively, S is the degree of saturation of the soil, K and K_s are the hydraulic conductivities of the soil at pressure head h and at saturation respectively and, α and χ are empirical constants determining the shape of the function.

The weak form of equation (9) with boundary condition is written as

$$\int_0^1 \delta C^T \frac{\partial}{\partial x}\left(\theta D\left(\frac{\partial C}{\partial x}\right)\right)\mathrm{d}x - \int_0^1 \delta C^T \frac{\partial}{\partial x}(uc)\,\mathrm{d}x - \int_0^1 \delta C^T \frac{\partial}{\partial t}(R\theta C)\,\mathrm{d}x \\ - \delta\lambda^T(C - C_0)\big|_{\Gamma_s} - \lambda^T \delta C\big|_{\Gamma_s} \tag{12}$$

where λ is a Lagrangian multiplier for enforcing the essential boundary condition and it is equal to flux term.

By using divergence theorem, and splitting the resultant equation, the final equation can be written as

$$\begin{aligned}\left[\mathbf{K}^{(1)}\right]\{C\}+\left[\mathbf{K}^{(2)}\right]\{C\}_{,t}+[G]\{\lambda\}=\{Q\}\\ \left[G^T\right]\{C\}=\{q\}\end{aligned} \tag{13}$$

where

$$\mathbf{K}_{IJ}^{(1)}=\int_0^1\left[\Phi_{I,x}^T\theta D\Phi_{J,x}+\Phi_I^T u\Phi_{J,x}\right]\mathrm{d}x\,, \tag{14a}$$

$$\mathbf{K}_{IJ}^{(2)}=\int_0^1\left[\Phi_I^T\theta R\Phi_J\right]\mathrm{d}x\,, \tag{14b}$$

$$\mathbf{G}_{IK}=\Phi_K\big|_{\Gamma_s I}\,, \tag{14c}$$

$$\mathbf{Q}_I=\Phi_I\theta Dg\big|_{\Gamma_E}\,, \tag{14d}$$

$$\mathbf{q}_K=C_{0K}\,. \tag{14e}$$

Using Crank–Nicholson method for time approximation, equation (13) can be written as

$$\begin{bmatrix}\mathbf{K}^{(1)*}+\mathbf{K}^{(2)} & G\\ G^T & 0\end{bmatrix}\begin{Bmatrix}C_n\\ \lambda\end{Bmatrix}=\begin{Bmatrix}R_n\\ q\end{Bmatrix} \tag{15a}$$

where

$$\begin{aligned}R_n&=\left(\left[\mathbf{K}^{(2)}\right]-(1-\varepsilon)\Delta t\left[\mathbf{K}^{(1)}\right]\right)\{C\}_{n-1}+\varepsilon\Delta t\{Q\}_n+(1-\varepsilon)\Delta t\{Q\}_{n-1}\,,\\ \mathbf{K}^{(1)*}&=\varepsilon\Delta t\left[\mathbf{K}^{(1)}\right]\end{aligned} \tag{15b}$$

where ε is a constant varying between 0 and 1, C_n and C_{n-1} are the nodal concentrations at start and end of time increment and, Q_n and Q_{n-1} are the nodal mass fluxes at start and end of time increment.

4 Numerical Examples: Results

The present model is applied to the problem considered by Diaw et al. (2001) for validation. The results of EFGM are compared with the results of Wamos – T model for two grid Peclet numbers: $P_e = 2$ and 50. In the present paper, central finite difference scheme ($\varepsilon = 0.5$) is used for time integration.

The parameters that are considered for the analysis are tabulated in Table 1. The total time of simulation is 12 hours. Comparison between the EFGM and Wamos – T results for both the advection – dispersion and advection – dispersion – sorption cases are presented in Figs. 1 and 2. It is seen from the figures that both the models provided similar results.

Table 1. Data for the 1-D analysis of contaminant equation

Parameters	Values
Initial condition for flow (cm)	−300.0
Boundary condition for flow at upper surface (cm)	−75.0
Boundary condition for flow at lower surface (cm)	−300.0
Saturated volumetric water content	0.368
Residual volumetric water content	0.102
Saturated hydraulic conductivity of soil (cm/s)	0.00922
α (cm^{-1})	0.0335
χ	2.0
Total duration of simulation (hours)	12.0
Length of the reach (cm)	50.0
Initial concentration (µg/cm^3)	0.0
Concentration at source boundary (µg/cm^3)	1.0

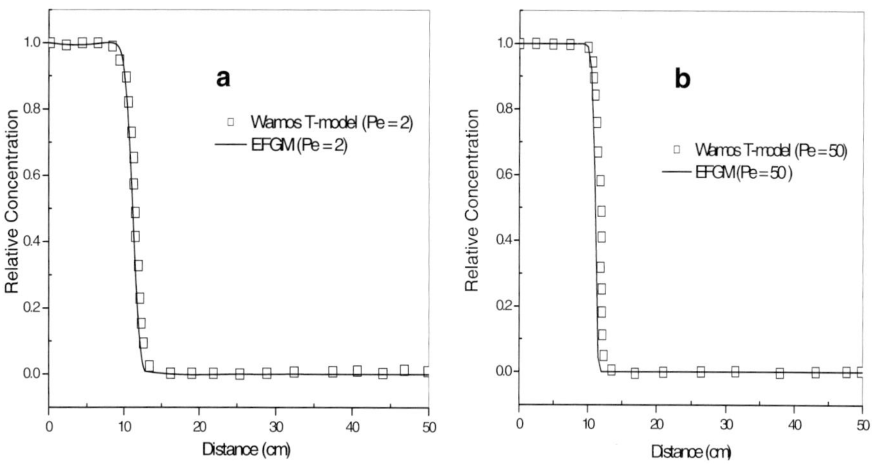

Fig. 1. Normalised concentration profile after 12 hrs for advection – dispersion

5 Conclusions

A truly meshless EFGM is obtained for one dimensional contaminant transport modelling for the unsaturated porous media. The present study has compared the results of EFGM with the Wamos – T model for two types of one dimensional contaminant transport processes that occur in the unsaturated porous media. It is noted that the EFGM generates excellent results for transport processes.

From application viewpoint, the implementation of the EFGM is very simple and straight forward, irrespective of the dimension of the problem and the shape of the domain under consideration. Further investigations are being

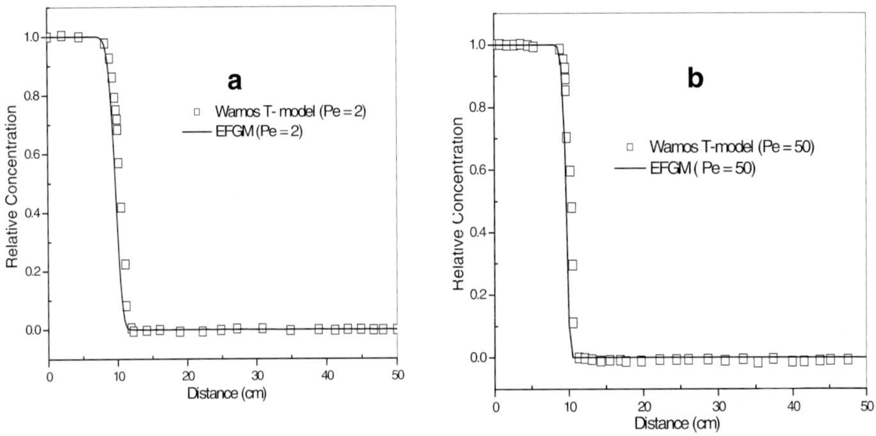

Fig. 2. Normalised concentration profile after 12 hrs for advection – dispersion – sorption

pursued for 2-D and 3-D contaminant transport modelling through the unsaturated porous media.

References

Belytschko T, Lu YY, Gu L (1994) Element-Free Galerkin Methods, Int J Num Meth Eng 37:229–256

Diaw EB, Lehmann F, Ackerer Ph (2001) Modélisation du transport d'un soluté réactif en milieux poreux non saturés: Modeling a non-conservative solute transfer in unsaturated porous media, Comptes Rendus de l'Académie des Sciences – Series II A – Earth and Planetary Science 333:129–132

van Genuchten MTh (1980) A closed-formed equation for predicting the hydraulic conductivity of unsaturated soils, Soil Sci Soc Am J 44:892–898

Lancaster P, Salkauskas K (1981) Surfaces Generated by Moving Least-Squares Methods, Math Comp 37:141–158

Rao BN, Rahman S (2000) An Efficient Meshless Method For Fracture Analysis of Cracks, Comp Mech 26:398–408

Index

SPRINGER PROCEEDINGS IN PHYSICS

66 **The Structure and Conformation of Amphiphilic Membranes**
Editors: R. Lipowsky, D. Richter, and K. Kremer

67 **Nonlinearity with Disorder**
Editors: F. Abdullaev, A.R. Bishop, and S. Pnevmatikos

68 **Time-Resolved Vibrational Spectroscopy V**
Editor: H. Takahashi

69 **Evolution of Dynamical Structures in Complex Systems**
Editors: R. Friedrich and A. Wunderlin

70 **Computational Approaches in Condensed-Matter Physics**
Editors: S. Miyashita, M. Imada, and H. Takayama

71 **Amorphous and Crystalline Silicon Carbide IV**
Editors: C.Y. Yang, M.M. Rahman, and G.L. Harris

72 **Computer Simulation Studies in Condensed-Matter Physics IV**
Editors: D.P. Landau, K.K. Mon, and H.-B. Schüttler

73 **Surface Science**
Principles and Applications
Editors: R.F. Howe, R.N: Lamb, and K. Wandelt

74 **Time-Resolved Vibrational Spectroscopy VI**
Editors: A. Lau, F. Siebert, and W. Werncke

75 **Computer Simulation Studies in Condensed-Matter Physics V**
Editors: D.P. Landau, K.K. Mon, and H.-B. Schüttler

76 **Computer Simulation Studies in Condensed-Matter Physics VI**
Editors: D.P. Landau, K.K. Mon, and H.-B. Schüttler

77 **Quantum Optics VI**
Editors: D.F. Walls and J.D. Harvey

78 **Computer Simulation Studies in Condensed-Matter Physics VII**
Editors: D.P. Landau, K.K. Mon, and H.-B. Schüttler

79 **Nonlinear Dynamics and Pattern Formation in Semiconductors and Devices**
Editor: F.-J. Niedernostheide

80 **Computer Simulation Studies in Condensed-Matter Physics VIII**
Editors: D.P. Landau, K.K. Mon, and H.-B. Schüttler

81 **Materials and Measurements in Molecular Electronics**
Editors: K. Kajimura and S. Kuroda

82 **Computer Simulation Studies in Condensed-Matter Physics IX**
Editors: D.P. Landau, K.K. Mon, and H.-B. Schüttler

83 **Computer Simulation Studies in Condensed-Matter Physics X**
Editors: D.P. Landau, K.K. Mon, and H.-B. Schüttler

84 **Computer Simulation Studies in Condensed-Matter Physics XI**
Editors: D.P. Landau and H.-B. Schüttler

85 **Computer Simulation Studies in Condensed-Matter Physics XII**
Editors: D.P. Landau, S.P. Lewis, and H.-B. Schüttler

86 **Computer Simulation Studies in Condensed-Matter Physics XIII**
Editors: D.P. Landau, S.P. Lewis, and H.-B. Schüttler

87 **Proceedings of the 25th International Conference on the Physics of Semiconductors**
Editors: N. Miura and T. Ando

88 **Starburst Galaxies**
Near and Far
Editors: L. Tacconi and D. Lutz

89 **Computer Simulation Studies in Condensed-Matter Physics XIV**
Editors: D.P. Landau, S.P. Lewis, and H.-B. Schüttler

90 **Computer Simulation Studies in Condensed-Matter Physics XV**
Editors: D.P. Landau, S.P. Lewis, and H.-B. Schüttler

Printing: Krips bv, Meppel
Binding: Stürtz, Würzburg